ТАБЛИЦЫ И НОМОГРАММЫ ДЛЯ РАСЧЕТА РЕЗУЛЬТАТОВ ГИДРОХИМИЧЕСКИХ АНАЛИЗОВ

TABLITSY I NOMOGRAMMY DLYA RASCHETA REZUL'TATOV GIDROKHIMICHESKIKH ANALIZOV

TABLES AND NOMOGRAMS OF HYDROCHEMICAL ANALYSIS

TABLES AND NOMOGRAMS OF HYDROCHEMICAL ANALYSIS

I. Yu. Sokolov

TRANSLATED FROM RUSSIAN

Springer Science+Business Media, LLC 1960

The original Russian text was published by the State Scientific and Technical Press for Literature on Geology and Conservation of Mineral Resources, Moscow, 1958 (under the title "Tables and Nomograms for the Calculation of the Results of Hydrochemical Analyses").

Library of Congress Catalog Card Number: 60-13952

ISBN 978-1-4899-4862-5   ISBN 978-1-4899-4860-1 (eBook)
DOI 10.1007/978-1-4899-4860-1

© 1960 Springer Science+Business Media New York
Originally published by Consultants Bureau Enterprises, Inc. in 1960.

CONTENTS

INTRODUCTION

The results of chemical analyses of water are expressed in three forms: weights, equivalents, and percent-equivalents. In addition, each of these forms has several variations. Thus, for example, when weights are used the salt content of water may be expressed in percent (%), parts per thousand ($^0/_{00}$), grams per liter (g/liter), milligrams per liter (mg/liter), milligrams per cubic meter (mg/m³), etc. The mineral content of water may be calculated as ions, oxides, salts, etc. When equivalents are used the results are given in gram-equivalents per liter or kilogram of water (g-equiv/liter or g-equiv/kg), milligram-equivalents per liter (meq/liter), etc. In the calculation of percent-equivalents, the sum of the equivalents of cations and anions are adjusted to 50 and 100%.

In generalizing and comparing hydrochemical analysis results obtained at different times by different laboratories, one must inevitably convert these results so that they conform to one system.

On the other hand, after completing the experimental part of an analysis, a hydrochemist must calculate the analytical results in two forms (weights and equivalents) or, more often, three (weights, equivalents, and percent-equivalents), and compare certain experimental data with theoretical values (for example, for aggressive carbon dioxide, the relation between the free carbon dioxide, bicarbonate ion, and pH, etc.).

These calculations are simplified considerably by using the tables and nomograms given in this book.

All the tables and nomograms are based on analytical results expressed in the form widely used in hydrogeological practice, namely, in milligrams per liter (weight form) and milligram-equivalents per liter (equivalent form). For calculation of the percent-equivalents, the sum of cation equivalents and the sum of anion equivalents were taken as 100% each.

The main section of this work consists of tables which make it possible to convert accurately the results of water analyses from one form into another, as follows:

1. German degrees of hardness into milligram-equivalents.
2. Milligram-equivalents of Ca^{2+}, Mg^{2+}, and HCO_3^- into German degrees of hardness.

3. Milligrams of ions into milligram-equivalents.
4. Milligram-equivalents of ions into milligrams.
5. Oxides into ions.
6. Nitrogen into the corresponding ions (NH_4^+, NO_2^-, and NO_3^-).
7. Ammonia (NH_3) into ammonium ion (NH_4^+).
8. Oxidizability with $KMnO_4$ to oxygen.

Tables 2, 21—38, 39—52, 53—55, 56, and 57 have been drawn up for the first time. Tables 1 and 3—20 are those of P. N. Butyrin and F. F. Laptev (see P. N. Butyrin and F. F. Laptev, Tables for Converting the Results of Water Analyses to the Equivalent Form [in Russian], Gosgeolizdat, 1933), which have been refined, supplemented, and recalculated in accordance with modern international atomic weights.

TABLES FOR CONVERTING WATER ANALYSIS RESULTS FROM ONE FORM TO ANOTHER

The tables make it possible to calculate the milligram-equivalents for any practically possible content of a component in water, accurate to the second decimal place, and the weight content of substances to tenths of a milligram (for low component concentrations the accuracy of the calculation may be increased to a further decimal place).

To reach this degree of accuracy in the calculations it is necessary to adhere to the rules presented in the following examples.

Example 1.

a) Given: 64.8 mg/liter of Cl^-. The Cl^- concentration in milligram-equivalents per liter is required.

In Table 12 we find the number 64 (whole milligrams) in the vertical column and move horizontally to the number 8 (tenths of a milligram) to read off the number 1.828 in the corresponding square. Rounding off this number to the second decimal place, we obtain the desired result: 1.83 meq/liter of Cl^-.

b) Given: 73.2 mg/liter of CaO. The Ca^{2+} concentration in milligrams per liter is required. In the corresponding square of Table 40 we find the number 52.32, and rounding this off to the first decimal place, we obtain the result: 52.3 mg/liter of Ca^{2+}.

Example 2.

a) Given: 664.8 mg/liter of Cl^-. The Cl^- concentration in milligram-equivalents per liter is required. The tenths of a milligram are neglected and we look for the milligram-equivalents for 664 mg. For this purpose the number 664 is divided by ten to give the value 66.4 mg and in the corresponding square of Table 12 we find the number 1.873. This figure is multiplied by ten to give the number of milligram-equivalents corresponding to 664 mg of Cl^-, i.e., 18.73 meq of Cl^-. To the value obtained we add the number of milligram-equivalents corresponding to 0.8 mg of Cl^-. This number is found in the first line of the table and equals 0.02. Thus, the final result equals $18.73 + 0.02 = 18.75$ meq/liter of Cl^-.

b) Given: 732.2 mg/liter of CaO. The Ca^{2+} concentration in milligrams per liter is required. As in Example 2a, from Table 40 we find that 73.2 mg of CaO corresponds to 52.32 mg of Ca^{2+} and, consequently, 732 mg of CaO corresponds to 523.2 mg of Ca^{2+}. By adding to this value 0.1 mg of Ca^{2+}, corresponding to 0.2 mg of CaO, we obtain the final result, 523.3 mg/liter of Ca^{2+}.

Example 3.

a) Given: 6664.8 mg/liter of Cl^-. The Cl^- concentration in milligram-equivalents per liter is required. From Table 12 we find the number of milligram-equivalents corresponding to 664.8 mg/liter of Cl^-, as described in Example 2a, and to the value obtained (18.75 meq) we add the number of milligram-equivalents corresponding to 6000 mg of Cl^-*, namely, 169.22 meq/liter. The final result equals $169.22 + 18.75 = 187.97$ meq/liter of Cl^-.

b) Given: 1732.2 mg/liter of CaO. The Ca^{2+} concentration in milligrams per liter is required. The number of milligrams of Ca^{2+} corresponding to 732.2 mg of CaO is found (as shown in Example 2b) and to the value obtained is added the number of milligrams of Ca^{2+} corresponding to 1000 mg of CaO. The final result equals $714.7 + 523.3 = 1238.0$ mg/liter of Ca^{2+}.

---

* The figures for thousands of milligrams of ions are given separately at the bottom of the table.

Table 1

I. Table for Converting Hardness, Expressed in German Degrees into Milligram-Equivalents

Whole degrees	Tenths of a degree									
	0	1	2	3	4	5	6	7	8	9
0	—	0,04	0,07	0,11	0,14	0,18	0,21	0,25	0,29	0,32
1	0,357	0,392	0,428	0,464	0,499	0,535	0,571	0,606	0,642	0,678
2	0,713	0,749	0,785	0,820	0,856	0,892	0,927	0,963	0,999	1,034
3	1,070	1,106	1,141	1,177	1,213	1,248	1,284	1,319	1,355	1,391
4	1,427	1,462	1,498	1,533	1,569	1,605	1,640	1,676	1,712	1,747
5	1,783	1,819	1,854	1,890	1,926	1,961	1,997	2,033	2,068	2,104
6	2,140	2,175	2,211	2,247	2,282	2,318	2,354	2,389	2,425	2,461
7	2,496	2,532	2,568	2,603	2,639	2,675	2,710	2,746	2,782	2,817
8	2,853	2,889	2,924	2,960	2,996	3,031	3,067	3,103	3,138	3,174
9	3,210	3,245	3,281	3,317	3,352	3,388	3,424	3,459	3,495	3,531
10	3,566	3,602	3,638	3,673	3,709	3,745	3,780	3,816	3,852	3,887
11	3,923	3,959	3,994	4,030	4,066	4,101	4,137	4,173	4,208	4,244
12	4,280	4,315	4,351	4,387	4,422	4,458	4,494	4,529	4,565	4,601
13	4,636	4,672	4,708	4,743	4,779	4,815	4,850	4,886	4,922	4,957
14	4,993	5,029	5,064	5,100	5,136	5,171	5,207	5,243	5,278	5,314
15	5,349	5,385	5,421	5,456	5,492	5,528	5,563	5,599	5,635	5,670
16	5,706	5,742	5,777	5,813	5,849	5,884	5,920	5,956	5,991	6,027
17	6,063	6,098	6,134	6,170	6,205	6,241	6,277	6,312	6,348	6,384
18	6,419	6,455	6,491	6,526	6,562	6,598	6,633	6,669	6,705	6,740
19	6,776	6,812	6,847	6,883	6,919	6,954	6,990	7,026	7,061	7,097
20	7,133	7,168	7,204	7,240	7,275	7,311	7,347	7,382	7,418	7,454
21	7,489	7,525	7,561	7,596	7,632	7,668	7,703	7,739	7,775	7,810
22	7,846	7,882	7,917	7,953	7,989	8,024	8,060	8,096	8,131	8,167
23	8,203	8,238	8,274	8,310	8,345	8,381	8,417	8,452	8,488	8,524
24	8,559	8,595	8,631	8,666	8,702	8,738	8,773	8,809	8,844	8,880
25	8,916	8,951	8,987	9,023	9,058	9,094	9,130	9,165	9,201	9,237
26	9,272	9,308	9,344	9,379	9,415	9,451	9,486	9,522	9,558	9,593
27	9,629	9,665	9,700	9,736	9,772	9,807	9,843	9,879	9,914	9,950
28	9,986	10,021	10,057	10,093	10,128	10,164	10,200	10,235	10,271	10,307
29	10,342	10,378	10,414	10,449	10,485	10,521	10,556	10,592	10,628	10,663
30	10,699	10,735	10,770	10,806	10,842	10,877	10,913	10,949	10,984	11,020
31	11,056	11,091	11,127	11,163	11,198	11,234	11,270	11,305	11,341	11,377
32	11,412	11,448	11,484	11,519	11,555	11,591	11,626	11,662	11,698	11,733
33	11,769	11,805	11,840	11,876	11,912	11,947	11,983	12,019	12,054	12,090
34	12,126	12,161	12,197	12,233	12,268	12,304	12,340	12,375	12,411	12,446
35	12,482	12,518	12,553	12,589	12,625	12,660	12,696	12,732	12,767	12,803
36	12,839	12,874	12,910	12,946	12,981	13,017	13,053	13,088	13,124	13,160
37	13,195	13,231	13,267	13,302	13,338	13,374	13,409	13,445	13,481	13,516
38	13,552	13,588	13,623	13,659	13,695	13,730	13,766	13,802	13,837	13,873
39	13,909	13,944	13,980	14,016	14,051	14,087	14,123	14,158	14,194	14,230
40	14,265	14,301	14,337	14,372	14,408	14,444	14,479	14,515	14,551	14,586
41	14,622	14,658	14,693	14,729	14,765	14,800	14,836	14,872	14,907	14,943
42	14,979	15,014	15,050	15,086	15,121	15,157	15,193	15,228	15,264	15,300
43	15,335	15,371	15,407	15,442	15,478	15,514	15,549	15,585	15,621	15,656
44	15,692	15,728	15,763	15,799	15,835	15,870	15,906	15,941	15,977	16,013
45	16,048	16,084	16,120	16,155	16,191	16,227	16,262	16,298	16,334	16,369
46	16,405	16,441	16,476	16,512	16,548	16,583	16,619	16,655	16,690	16,726
47	16,762	16,797	16,833	16,869	16,904	16,940	16,976	17,011	17,047	17,083
48	17,118	17,154	17,190	17,225	17,261	17,297	17,332	17,368	17,404	17,439
49	17,475	17,511	17,546	17,582	17,618	17,653	17,689	17,725	17,760	17,796
50	17,832	17,867	17,903	17,939	17,974	18,010	18,046	18,081	18,117	18,153

Whole degrees	0	1	2	3	4	5	6	7	8	9
					Tenths of a degree					
51	18,188	18,224	18,260	18,295	18,331	18,367	18,402	18,438	18,474	18,509
52	18,545	18,581	18,616	18,652	18,688	18,723	18,759	18,795	18,830	18,866
53	18,902	18,937	18,973	19,009	19,044	19,080	19,116	19,151	19,187	19,223
54	19,258	19,294	19,330	19,365	19,401	19,436	19,472	19,508	19,543	19,579
55	19,615	19,650	19,686	19,722	19,757	19,793	19,829	19,864	19,900	19,936
56	19,971	20,007	20,043	20,078	20,114	20,150	20,185	20,221	20,257	20,292
57	20,328	20,364	20,399	20,435	20,471	20,506	20,542	20,578	20,613	20,649
58	20,685	20,720	20,756	20,792	20,827	20,863	20,899	20,934	20,970	21,006
59	21,041	21,077	21,113	21,148	21,184	21,220	21,255	21,291	21,327	21,362
60	21,398	21,434	21,469	21,505	21,541	21,576	21,612	21,648	21,683	21,719
61	21,755	21,790	21,826	21,862	21,897	21,933	21,969	22,004	22,040	22,076
62	22,111	22,147	22,183	22,218	22,254	22,290	22,325	22,361	22,397	22,432
63	22,468	22,504	22,539	22,575	22,611	22,646	22,682	22,718	22,753	22,789
64	22,825	22,860	22,896	22,932	22,967	23,003	23,038	23,074	23,110	23,145
65	23,181	23,217	23,254	23,288	23,324	23,359	23,395	23,431	23,466	23,502
66	23,538	23,573	23,609	23,645	23,680	23,716	23,752	23,787	23,823	23,859
67	23,894	23,930	23,966	24,001	24,037	24,073	24,108	24,144	24,180	24,215
68	24,251	24,287	24,322	24,358	24,394	24,429	24,465	24,501	24,536	24,572
69	24,608	24,643	24,679	24,715	24,750	24,786	24,822	24,857	24,893	24,929
70	24,964	25,000	25,036	25,071	25,107	25,143	25,178	25,214	25,250	25,285
71	25,321	25,357	25,392	25,428	25,464	25,499	25,535	25,571	25,606	25,642
72	25,678	25,713	25,749	25,785	25,820	25,856	25,892	25,927	25,963	25,999
73	26,034	26,070	26,106	26,141	26,177	26,213	26,248	26,284	26,320	26,355
74	26,391	26,427	26,462	26,498	26,533	26,569	26,605	26,640	26,676	26,712
75	26,747	26,783	26,819	26,854	26,890	26,926	26,961	26,997	27,033	27,068
76	27,104	27,140	27,175	27,211	27,247	27,282	27,318	27,354	27,389	27,425
77	27,461	27,496	27,532	27,568	27,603	27,639	27,675	27,710	27,746	27,782
78	27,817	27,853	27,889	27,924	27,960	27,996	28,031	28,067	28,103	28,138
79	28,174	28,210	28,245	28,281	28,317	28,352	28,388	28,424	28,459	28,495
80	28,531	28,566	28,602	28,638	28,673	28,709	28,745	28,780	28,816	28,852
81	28,887	28,923	28,959	28,994	29,030	29,066	29,101	29,137	29,173	29,208
82	29,244	29,280	29,315	29,351	29,387	29,422	29,458	29,494	29,529	29,565
83	29,601	29,636	29,672	29,708	29,743	29,779	29,815	29,850	29,886	29,922
84	29,957	29,993	30,028	30,064	30,100	30,135	30,171	30,207	30,242	30,278
85	30,314	30,349	30,385	30,421	30,456	30,492	30,528	30,563	30,599	30,635
86	30,670	30,706	30,742	30,777	30,813	30,849	30,884	30,920	30,956	30,991
87	31,027	31,063	31,098	31,134	31,170	31,205	31,241	31,277	31,312	31,348
88	31,384	31,419	31,455	31,491	31,526	31,562	31,598	31,633	31,669	31,705
89	31,740	31,776	31,812	31,847	31,883	31,919	31,954	31,990	32,026	32,061
90	32,097	32,133	32,168	32,204	32,240	32,275	32,311	32,347	32,382	32,418
91	32,454	32,489	32,525	32,561	32,596	32,632	32,668	32,703	32,739	32,775
92	32,810	32,846	32,882	32,917	32,953	32,989	33,024	33,060	33,096	33,131
93	33,167	33,203	33,238	33,274	33,310	33,345	33,381	33,417	33,452	33,488
94	33,524	33,559	33,595	33,630	33,666	33,702	33,737	33,773	33,809	33,844
95	33,880	33,916	33,951	33,987	34,023	34,058	34,094	34,130	34,165	34,201
96	34,237	34,272	34,308	34,344	34,379	34,415	34,451	34,486	34,522	34,558
97	34,593	34,629	34,665	34,700	34,736	34,772	34,807	34,843	34,879	34,914
98	34,950	34,986	35,021	35,057	35,093	35,128	35,164	35,200	35,235	35,271
99	35,307	35,342	35,378	35,414	35,449	35,485	35,521	35,556	35,592	35,628
100	35,663	35,699	35,735	35,770	35,806	35,842	35,877	35,913	35,949	35,982

Hardness in German degrees	1000	2000	3000	4000	5000	6000
meq	356,63	713,27	1069,90	1426,53	1783,16	2139,80

Table 2

II. Table for Converting Milligram-Equivalents of Ca^{2+}, Mg^{2+}, and HCO_3^-
into German Degrees of Hardness

Whole and tenths of meq	Hundredths of a meq									
	0	1	2	3	4	5	6	7	8	9
0,0	—	0,03	0,06	0,08	0,11	0,14	0,17	0,20	0,22	0,25
0,1	0,28	0,31	0,34	0,36	0,39	0,42	0,45	0,48	0,50	0,53
0,2	0,56	0,59	0,62	0,64	0,67	0,70	0,73	0,76	0,79	0,81
0,3	0,84	0,87	0,90	0,93	0,95	0,98	1,01	1,04	1,07	1,09
0,4	1,12	1,15	1,18	1,21	1,23	1,26	1,29	1,32	1,35	1,37
0,5	1,40	1,43	1,46	1,49	1,51	1,54	1,57	1,60	1,63	1,65
0,6	1,68	1,71	1,74	1,77	1,79	1,82	1,85	1,88	1,91	1,93
0,7	1,96	1,99	2,02	2,05	2,07	2,10	2,13	2,16	2,19	2,22
0,8	2,24	2,27	2,30	2,33	2,36	2,38	2,41	2,44	2,47	2,50
0,9	2,52	2,55	2,58	2,61	2,64	2,66	2,69	2,72	2,75	2,78
1,0	2,80	2,83	2,86	2,89	2,92	2,94	2,97	3,00	3,03	3,06
1,1	3,08	3,11	3,14	3,17	3,20	3,22	3,25	3,28	3,31	3,34
1,2	3,36	3,39	3,42	3,45	3,48	3,50	3,53	3,56	3,59	3,62
1,3	3,65	3,67	3,70	3,73	3,76	3,79	3,81	3,84	3,87	3,90
1,4	3,93	3,95	3,98	4,01	4,04	4,07	4,09	4,12	4,15	4,18
1,5	4,21	4,23	4,26	4,29	4,32	4,35	4,37	4,40	4,43	4,46
1,6	4,49	4,51	4,54	4,57	4,60	4,63	4,65	4,68	4,71	4,74
1,7	4,77	4,79	4,82	4,85	4,88	4,91	4,94	4,96	4,99	5,02
1,8	5,05	5,08	5,10	5,13	5,16	5,19	5,22	5,24	5,27	5,30
1,9	5,33	5,36	5,38	5,41	5,44	5,47	5,50	5,52	5,55	5,58
2,0	5,61	5,64	5,66	5,69	5,72	5,75	5,78	5,80	5,83	5,86
2,1	5,89	5,92	5,94	5,97	6,00	6,03	6,06	6,08	6,11	6,14
2,2	6,17	6,20	6,22	6,25	6,28	6,31	6,34	6,37	6,39	6,42
2,3	6,45	6,48	6,51	6,53	6,56	6,59	6,62	6,65	6,67	6,70
2,4	6,73	6,76	6,79	6,81	6,84	6,87	6,90	6,93	6,95	6,98
2,5	7,01	7,04	7,07	7,09	7,12	7,15	7,18	7,21	7,23	7,26
2,6	7,29	7,32	7,35	7,37	7,40	7,43	7,46	7,49	7,51	7,54
2,7	7,57	7,60	7,63	7,65	7,68	7,71	7,74	7,77	7,80	7,82
2,8	7,85	7,88	7,91	7,94	7,96	7,99	8,02	8,05	8,08	8,10
2,9	8,13	8,16	8,19	8,22	8,24	8,27	8,30	8,33	8,36	8,38
3,0	8,41	8,44	8,47	8,50	8,52	8,55	8,58	8,61	8,64	8,66
3,1	8,69	8,72	8,75	8,78	8,80	8,83	8,86	8,89	8,92	8,94
3,2	8,97	9,00	9,03	9,06	9,08	9,11	9,14	9,17	9,20	9,23
3,3	9,25	9,28	9,31	9,34	9,37	9,39	9,42	9,45	9,48	9,51
3,4	9,53	9,56	9,59	9,62	9,65	9,67	9,70	9,73	9,76	9,79
3,5	9,81	9,84	9,87	9,90	9,93	9,95	9,98	10,01	10,04	10,07
3,6	10,09	10,12	10,15	10,18	10,21	10,23	10,26	10,29	10,32	10,35
3,7	10,37	10,40	10,43	10,46	10,49	10,51	10,54	10,57	10,60	10,63
3,8	10,66	10,68	10,71	10,74	10,77	10,80	10,82	10,85	10,88	10,91
3,9	10,94	10,96	10,99	11,02	11,05	11,08	11,10	11,13	11,16	11,19
4,0	11,22	11,24	11,27	11,30	11,33	11,36	11,38	11,41	11,44	11,47
4,1	11,50	11,52	11,55	11,58	11,61	11,64	11,66	11,69	11,72	11,75
4,2	11,78	11,80	11,83	11,86	11,89	11,92	11,95	11,97	12,00	12,03
4,3	12,06	12,09	12,11	12,14	12,17	12,20	12,23	12,25	12,28	12,31
4,4	12,34	12,37	12,39	12,42	12,45	12,48	12,51	12,53	12,56	12,59
4,5	12,62	12,65	12,67	12,70	12,73	12,76	12,79	12,81	12,84	12,87
4,6	12,90	12,93	12,95	12,98	13,01	13,04	13,07	13,09	13,12	13,15
4,7	13,18	13,21	13,23	13,26	13,29	13,32	13,35	13,38	13,40	13,43
4,8	13,46	13,49	13,52	13,54	13,57	13,60	13,63	13,66	13,68	13,71
4,9	13,74	13,77	13,80	13,82	13,85	13,88	13,91	13,94	13,96	13,99
5,0	14,02	14,05	14,08	14,10	14,13	14,16	14,19	14,22	14,24	14,27

Whole and tenths of meq	Hundredths of a meq									
	0	1	2	3	4	5	6	7	8	9
5,1	14,30	14,33	14,36	14,38	14,41	14,44	14,47	14,50	14,52	14,55
5,2	14,58	14,61	14,64	14,66	14,69	14,72	14,75	14,78	14,81	14,83
5,3	14,86	14,89	14,92	14,95	14,97	15,00	15,03	15,06	15,09	15,11
5,4	15,14	15,17	15,20	15,23	15,25	15,28	15,31	15,34	15,37	15,39
5,5	15,42	15,45	15,48	15,51	15,53	15,56	15,59	15,62	15,65	15,67
5,6	15,70	15,73	15,76	15,79	15,81	15,84	15,87	15,90	15,93	15,95
5,7	15,98	16,01	16,04	16,07	16,09	16,12	16,15	16,18	16,21	16,24
5,8	16,26	16,29	16,32	16,35	16,38	16,40	16,43	16,46	16,49	16,52
5,9	16,54	16,57	16,60	16,63	16,66	16,68	16,71	16,74	16,77	16,80
6,0	16,82	16,85	16,88	16,91	16,94	16,96	16,99	17,02	17,05	17,08
6,1	17,10	17,13	17,16	17,19	17,22	17,24	17,27	17,30	17,33	17,36
6,2	17,38	17,41	17,44	17,47	17,50	17,52	17,55	17,58	17,61	17,64
6,3	17,67	17,69	17,72	17,75	17,78	17,81	17,83	17,86	17,89	17,92
6,4	17,95	17,97	18,00	18,03	18,06	18,09	18,11	18,14	18,17	18,20
6,5	18,23	18,25	18,28	18,31	18,34	18,37	18,39	18,42	18,45	18,48
6,6	18,51	18,53	18,56	18,59	18,62	18,65	18,67	18,70	18,73	18,76
6,7	18,79	18,81	18,84	18,87	18,90	18,93	18,96	18,98	19,01	19,04
6,8	19,07	19,10	19,12	19,15	19,18	19,21	19,24	19,26	19,29	19,32
6,9	19,35	19,38	19,40	19,43	19,46	19,49	19,52	19,54	19,57	19,60
7,0	19,63	19,66	19,68	19,71	19,74	19,77	19,80	19,82	19,85	19,88
7,1	19,91	19,94	19,96	19,99	20,02	20,05	20,08	20,10	20,13	20,16
7,2	20,19	20,22	20,24	20,27	20,30	20,33	20,36	20,39	20,41	20,44
7,3	20,47	20,50	20,53	20,55	20,58	20,61	20,64	20,67	20,69	20,72
7,4	20,75	20,78	20,81	20,83	20,86	20,89	20,92	20,95	20,97	21,00
7,5	21,03	21,06	21,09	21,11	21,14	21,17	21,20	21,23	21,25	21,28
7,6	21,31	21,34	21,37	21,39	21,42	21,45	21,48	21,51	21,53	21,56
7,7	21,59	21,62	21,65	21,67	21,70	21,73	21,76	21,79	21,82	21,84
7,8	21,87	21,90	21,93	21,95	21,98	22,01	22,04	22,07	22,10	22,12
7,9	22,15	22,18	22,21	22,24	22,26	22,29	22,32	22,35	22,38	22,40
8,0	22,43	22,46	22,49	22,52	22,54	22,57	22,60	22,63	22,66	22,68
8,1	22,71	22,74	22,77	22,80	22,82	22,85	22,88	22,91	22,94	22,96
8,2	22,99	23,02	23,05	23,08	23,10	23,13	23,16	23,19	23,22	23,25
8,3	23,27	23,30	23,33	23,36	23,39	23,41	23,44	23,47	23,50	23,53
8,4	23,55	23,58	23,61	23,64	23,67	23,69	23,72	23,75	23,78	23,81
8,5	23,83	23,86	23,89	23,92	23,95	23,97	24,00	24,03	24,06	24,09
8,6	24,11	24,14	24,17	24,20	24,23	24,25	24,28	24,31	24,34	24,37
8,7	24,39	24,42	24,45	24,48	24,51	24,53	24,56	24,59	24,62	24,65
8,8	24,68	24,70	24,73	24,76	24,79	24,82	24,84	24,87	24,90	24,93
8,9	24,96	24,98	25,01	25,04	25,07	25,10	25,12	25,15	25,18	25,21
9,0	25,24	25,26	25,29	25,32	25,35	25,38	25,40	25,43	25,46	25,49
9,1	25,52	25,54	25,57	25,60	25,63	25,66	25,68	25,71	25,74	25,77
9,2	25,80	25,82	25,85	25,88	25,91	25,94	25,97	25,99	26,02	26,05
9,3	26,08	26,11	26,13	26,16	26,19	26,22	26,25	26,27	26,30	26,33
9,4	26,36	26,39	26,41	26,44	26,47	26,50	26,53	26,55	26,58	26,61
9,5	26,64	26,67	26,69	26,72	26,75	26,78	26,81	26,83	26,86	26,89
9,6	26,92	26,95	26,97	27,00	27,03	27,06	27,09	27,11	27,14	27,17
9,7	27,20	27,23	27,25	27,28	27,31	27,34	27,37	27,40	27,42	27,45
9,8	27,48	27,51	27,54	27,56	27,59	27,62	27,65	27,68	27,70	27,73
9,9	27,76	27,79	27,82	27,84	27,87	27,90	27,93	27,96	27,98	28,01
10,0	28,04	28,07	28,10	28,12	28,15	28,18	28,21	28,24	28,26	28,29

meq	100	200	300	400	500	600	700	800	900	1000
Hardness in German degrees	280,4	560,8	841,2	1121,6	1402,0	1682,4	1962,8	2243,2	2523,6	2804,0

III. Tables for Converting Milligrams into Milligram-Equivalents

Table 3

Conversion of Milligrams of Na$^+$ into Milligram-Equivalents
(equivalent weight of Na$^+$ = 22.991)

Whole mg	Tenths of a mg									
	0	1	2	3	4	5	6	7	8	9
0	—	0,00	0,01	0,01	0,02	0,02	0,03	0,03	0,03	0,04
1	0,043	0,048	0,052	0,057	0,061	0,065	0,070	0,074	0,078	0,083
2	0,087	0,091	0,096	0,100	0,104	0,109	0,113	0,117	0,122	0,126
3	0,130	0,135	0,139	0,144	0,148	0,152	0,157	0,161	0,165	0,170
4	0,174	0,178	0,183	0,187	0,191	0,196	0,200	0,204	0,209	0,213
5	0,217	0,222	0,226	0,231	0,235	0,239	0,244	0,248	0,252	0,257
6	0,261	0,265	0,270	0,274	0,278	0,283	0,287	0,291	0,296	0,300
7	0,304	0,309	0,313	0,318	0,322	0,326	0,331	0,335	0,339	0,344
8	0,348	0,352	0,357	0,361	0,365	0,370	0,374	0,378	0,383	0,387
9	0,391	0,396	0,400	0,405	0,409	0,413	0,418	0,422	0,426	0,431
10	0,435	0,439	0,444	0,448	0,452	0,457	0,461	0,465	0,470	0,474
11	0,478	0,483	0,487	0,491	0,496	0,500	0,505	0,509	0,513	0,518
12	0,522	0,526	0,531	0,535	0,539	0,544	0,548	0,552	0,557	0,561
13	0,565	0,570	0,574	0,578	0,583	0,587	0,592	0,596	0,600	0,605
14	0,609	0,613	0,618	0,622	0,626	0,631	0,635	0,639	0,644	0,648
15	0,652	0,657	0,661	0,665	0,670	0,674	0,679	0,683	0,687	0,692
16	0,696	0,700	0,705	0,709	0,713	0,718	0,722	0,726	0,731	0,735
17	0,739	0,744	0,748	0,752	0,757	0,761	0,766	0,770	0,774	0,779
18	0,783	0,787	0,792	0,796	0,800	0,805	0,809	0,813	0,818	0,822
19	0,826	0,831	0,835	0,839	0,844	0,848	0,853	0,857	0,861	0,866
20	0,870	0,874	0,879	0,883	0,887	0,892	0,896	0,900	0,905	0,909
21	0,913	0,918	0,922	0,926	0,931	0,935	0,939	0,944	0,948	0,953
22	0,957	0,961	0,966	0,970	0,974	0,979	0,983	0,987	0,992	0,996
23	1,000	1,005	1,009	1,013	1,018	1,022	1,026	1,031	1,035	1,040
24	1,044	1,048	1,053	1,057	1,061	1,066	1,070	1,074	1,079	1,083
25	1,087	1,092	1,096	1,100	1,105	1,109	1,113	1,118	1,122	1,127
26	1,131	1,135	1,140	1,144	1,148	1,153	1,157	1,161	1,166	1,170
27	1,174	1,179	1,183	1,187	1,192	1,196	1,200	1,205	1,209	1,214
28	1,218	1,222	1,227	1,231	1,235	1,240	1,244	1,248	1,253	1,257
29	1,261	1,266	1,270	1,274	1,279	1,283	1,287	1,292	1,296	1,301
30	1,305	1,309	1,314	1,318	1,322	1,327	1,331	1,335	1,340	1,344
31	1,348	1,353	1,357	1,361	1,366	1,370	1,374	1,379	1,383	1,387
32	1,392	1,396	1,401	1,405	1,409	1,414	1,418	1,422	1,427	1,431
33	1,435	1,440	1,444	1,448	1,453	1,457	1,461	1,466	1,470	1,474
34	1,479	1,483	1,488	1,492	1,496	1,501	1,505	1,509	1,514	1,518
35	1,522	1,527	1,531	1,535	1,540	1,544	1,548	1,553	1,557	1,561
36	1,566	1,570	1,575	1,579	1,583	1,588	1,592	1,596	1,601	1,605
37	1,609	1,614	1,618	1,622	1,627	1,631	1,635	1,640	1,644	1,648
38	1,653	1,657	1,662	1,666	1,670	1,675	1,679	1,683	1,688	1,692
39	1,696	1,701	1,705	1,709	1,714	1,718	1,722	1,727	1,731	1,735
40	1,740	1,744	1,748	1,753	1,757	1,762	1,766	1,770	1,775	1,779
41	1,783	1,788	1,792	1,796	1,801	1,805	1,809	1,814	1,818	1,822
42	1,827	1,831	1,835	1,840	1,844	1,849	1,853	1,857	1,862	1,866
43	1,870	1,875	1,879	1,883	1,888	1,892	1,896	1,901	1,905	1,909
44	1,914	1,918	1,922	1,927	1,931	1,936	1,940	1,944	1,949	1,953
45	1,957	1,962	1,966	1,970	1,975	1,979	1,983	1,988	1,992	1,996
46	2,001	2,005	2,009	2,014	2,018	2,023	2,027	2,031	2,036	2,040
47	2,044	2,049	2,053	2,057	2,062	2,066	2,070	2,075	2,079	2,083
48	2,088	2,092	2,096	2,101	2,105	2,110	2,114	2,118	2,123	2,127
49	2,131	2,136	2,140	2,144	2,149	2,153	2,157	2,162	2,166	2,170
50	2,175	2,179	2,183	2,188	2,192	2,196	2,201	2,205	2,210	2,214

Whole mg	Tenths of a mg									
	0	1	2	3	4	5	6	7	8	9
51	2,218	2,223	2,227	2,231	2,236	2,240	2,244	2,249	2,253	2,257
52	2,262	2,266	2,270	2,275	2,279	2,283	2,288	2,292	2,297	2,301
53	2,305	2,310	2,314	2,318	2,323	2,327	2,331	2,336	2,340	2,344
54	2,349	2,353	2,357	2,362	2,366	2,370	2,375	2,379	2,384	2,388
55	2,392	2,397	2,401	2,405	2,410	2,414	2,418	2,423	2,427	2,431
56	2,436	2,440	2,444	2,449	2,453	2,457	2,462	2,466	2,471	2,475
57	2,479	2,484	2,488	2,492	2,497	2,501	2,505	2,510	2,514	2,518
58	2,523	2,527	2,531	2,536	2,540	2,544	2,549	2,553	2,558	2,562
59	2,566	2,571	2,575	2,579	2,584	2,588	2,592	2,597	2,601	2,605
60	2,610	2,614	2,618	2,623	2,627	2,631	2,636	2,640	2,645	2,649
61	2,653	2,658	2,662	2,666	2,671	2,675	2,679	2,684	2,688	2,692
62	2,697	2,701	2,705	2,710	2,714	2,718	2,723	2,727	2,731	2,736
63	2,740	2,745	2,749	2,753	2,758	2,762	2,766	2,771	2,775	2,779
64	2,784	2,788	2,792	2,797	2,801	2,805	2,810	2,814	2,818	2,823
65	2,827	2,832	2,836	2,840	2,845	2,849	2,853	2,858	2,862	2,866
66	2,871	2,875	2,879	2,884	2,888	2,892	2,897	2,901	2,905	2,910
67	2,914	2,919	2,923	2,927	2,932	2,936	2,940	2,945	2,949	2,953
68	2,958	2,962	2,966	2,971	2,975	2,979	2,984	2,988	2,992	2,997
69	3,001	3,006	3,010	3,014	3,019	3,023	3,027	3,032	3,036	3,040
70	3,045	3,049	3,053	3,058	3,062	3,066	3,071	3,075	3,079	3,084
71	3,088	3,092	3,097	3,101	3,106	3,110	3,114	3,119	3,123	3,127
72	3,132	3,136	3,140	3,145	3,149	3,153	3,158	3,162	3,166	3,171
73	3,175	3,179	3,184	3,188	3,193	3,197	3,201	3,206	3,210	3,214
74	3,219	3,223	3,227	3,232	3,236	3,240	3,245	3,249	3,253	3,258
75	3,262	3,266	3,271	3,275	3,280	3,284	3,288	3,293	3,297	3,301
76	3,306	3,310	3,314	3,319	3,323	3,327	3,332	3,336	3,340	3,345
77	3,349	3,353	3,358	3,362	3,367	3,371	3,375	3,380	3,384	3,388
78	3,393	3,397	3,401	3,406	3,410	3,414	3,419	3,423	3,427	3,432
79	3,436	3,440	3,445	3,449	3,454	3,458	3,462	3,467	3,471	3,475
80	3,480	3,484	3,488	3,493	3,497	3,501	3,506	3,510	3,514	3,519
81	3,523	3,527	3,532	3,536	3,540	3,545	3,549	3,554	3,558	3,562
82	3,567	3,571	3,575	3,580	3,584	3,588	3,593	3,597	3,601	3,606
83	3,610	3,614	3,619	3,623	3,627	3,632	3,636	3,641	3,645	3,649
84	3,654	3,658	3,662	3,667	3,671	3,675	3,680	3,684	3,688	3,693
85	3,697	3,701	3,706	3,710	3,714	3,719	3,723	3,728	3,732	3,736
86	3,741	3,745	3,749	3,754	3,758	3,762	3,767	3,771	3,775	3,780
87	3,784	3,788	3,793	3,797	3,801	3,806	3,810	3,815	3,819	3,823
88	3,828	3,832	3,836	3,841	3,845	3,849	3,854	3,858	3,862	3,867
89	3,871	3,875	3,880	3,884	3,888	3,893	3,897	3,902	3,906	3,910
90	3,915	3,919	3,923	3,928	3,932	3,936	3,941	3,945	3,949	3,954
91	3,958	3,962	3,967	3,971	3,975	3,980	3,984	3,988	3,993	3,997
92	4,002	4,006	4,010	4,015	4,019	4,023	4,028	4,032	4,036	4,041
93	4,045	4,049	4,054	4,058	4,062	4,067	4,071	4,075	4,080	4,084
94	4,089	4,093	4,097	4,102	4,106	4,110	4,115	4,119	4,123	4,128
95	4,132	4,136	4,141	4,145	4,149	4,154	4,158	4,162	4,167	4,171
96	4,176	4,180	4,184	4,189	4,193	4,197	4,202	4,206	4,210	4,215
97	4,219	4,223	4,228	4,232	4,236	4,241	4,245	4,249	4,254	4,258
98	4,263	4,267	4,271	4,276	4,280	4,284	4,289	4,293	4,297	4,302
99	4,306	4,310	4,315	4,319	4,323	4,328	4,332	4,336	4,341	4,345
100	4,350	4,354	4,358	4,363	4,367	4,371	4,376	4,380	4,384	4,388

mg	1000	2000	3000	4000	5000	6000	7000	8000	9000	10 000
meq	43,50	86,99	130,49	173,98	217,48	260,97	304,47	347,96	391.46	434,95

11

Table 4

Conversion of Milligrams of Ca^{2+} into Milligram-Equivalents
(equivalent weight of Ca^{2+} = 20.04)

Whole mg	Tenths of a mg									
	0	1	2	3	4	5	6	7	8	9
0	—	0,00	0,01	0,01	0,02	0,02	0,03	0,04	0,04	0,04
1	0,050	0,055	0,060	0,065	0,070	0,075	0,080	0,085	0,090	0,095
2	0,100	0,105	0,110	0,115	0,120	0,125	0,130	0,135	0,140	0,145
3	0,150	0,155	0,160	0,165	0,170	0,175	0,180	0,185	0,190	0,195
4	0,200	0,205	0,210	0,215	0,220	0,225	0,230	0,235	0,240	0,245
5	0,250	0,254	0,259	0,264	0,269	0,274	0,279	0,284	0,289	0,294
6	0,299	0,304	0,309	0,314	0,319	0,324	0,329	0,334	0,339	0,344
7	0,349	0,354	0,359	0,364	0,369	0,374	0,379	0,384	0,389	0,394
8	0,399	0,404	0,409	0,414	0,419	0,424	0,429	0,434	0,439	0,444
9	0,449	0,454	0,459	0,464	0,469	0,474	0,479	0,484	0,489	0,494
10	0,499	0,504	0,509	0,514	0,519	0,524	0,529	0,534	0,539	0,544
11	0,549	0,554	0,559	0,564	0,569	0,574	0,579	0,584	0,589	0,594
12	0,599	0,604	0,609	0,614	0,619	0,624	0,629	0,634	0,639	0,644
13	0,649	0,654	0,659	0,664	0,669	0,674	0,679	0,684	0,689	0,694
14	0,699	0,704	0,709	0,714	0,719	0,724	0,729	0,734	0,739	0,744
15	0,749	0,753	0,758	0,763	0,768	0,773	0,778	0,783	0,788	0,793
16	0,798	0,803	0,808	0,813	0,818	0,823	0,828	0,833	0,838	0,843
17	0,848	0,853	0,858	0,863	0,868	0,873	0,878	0,883	0,888	0,893
18	0,898	0,903	0,908	0,913	0,918	0,923	0,928	0,933	0,938	0,943
19	0,948	0,953	0,958	0,963	0,968	0,973	0,978	0,983	0,988	0,993
20	0,998	1,003	1,008	1,013	1,018	1,023	1,028	1,033	1,038	1,043
21	1,048	1,053	1,058	1,063	1,068	1,073	1,078	1,083	1,088	1,093
22	1,098	1,103	1,108	1,113	1,118	1,123	1,128	1,133	1,138	1,143
23	1,148	1,153	1,158	1,163	1,168	1,173	1,178	1,183	1,188	1,193
24	1,198	1,203	1,208	1,213	1,218	1,223	1,228	1,233	1,238	1,243
25	1,248	1,252	1,257	1,262	1,267	1,272	1,277	1,282	1,287	1,292
26	1,297	1,302	1,307	1,312	1,317	1,322	1,327	1,332	1,337	1,342
27	1,347	1,352	1,357	1,362	1,367	1,372	1,377	1,382	1,387	1,392
28	1,397	1,402	1,407	1,412	1,417	1,422	1,427	1,432	1,437	1,442
29	1,447	1,452	1,457	1,462	1,467	1,472	1,477	1,482	1,487	1,492
30	1,497	1,502	1,507	1,512	1,517	1,522	1,527	1,532	1,537	1,542
31	1,547	1,552	1,557	1,562	1,567	1,572	1,577	1,582	1,587	1,592
32	1,597	1,602	1,607	1,612	1,617	1,622	1,627	1,632	1,637	1,642
33	1,647	1,652	1,657	1,662	1,667	1,672	1,677	1,682	1,687	1,692
34	1,697	1,702	1,707	1,712	1,717	1,722	1,727	1,732	1,737	1,742
35	1,747	1,751	1,756	1,761	1,766	1,771	1,776	1,781	1,786	1,791
36	1,796	1,801	1,806	1,811	1,816	1,821	1,826	1,831	1,836	1,841
37	1,846	1,851	1,856	1,861	1,866	1,871	1,876	1,881	1,886	1,891
38	1,896	1,901	1,906	1,911	1,916	1,921	1,926	1,931	1,936	1,941
39	1,946	1,951	1,956	1,961	1,966	1,971	1,976	1,981	1,986	1,991
40	1,996	2,001	2,006	2,011	2,016	2,021	2,026	2,031	2,036	2,041
41	2,046	2,051	2,056	2,061	2,066	2,071	2,076	2,081	2,086	2,091
42	2,096	2,101	2,106	2,111	2,116	2,121	2,126	2,131	2,136	2,141
43	2,146	2,151	2,156	2,161	2,166	2,171	2,176	2,181	2,186	2,191
44	2,196	2,201	2,206	2,211	2,216	2,221	2,226	2,231	2,236	2,241
45	2,246	2,250	2,255	2,260	2,265	2,270	2,275	2,280	2,285	2,290
46	2,295	2,300	2,305	2,310	2,315	2,320	2,325	2,330	2,335	2,340
47	2,345	2,350	2,355	2,360	2,365	2,370	2,375	2,380	2,385	2,390
48	2,395	2,400	2,405	2,410	2,415	2,420	2,425	2,430	2,435	2,440
49	2,445	2,450	2,455	2,460	2,465	2,470	2,475	2,480	2,485	2,490
50	2,495	2,500	2,505	2,510	2,515	2,520	2,525	2,530	2,535	2,540

Whole mg	Tenths of a mg									
	0	1	2	3	4	5	6	7	8	9
51	2,545	2,550	2,555	2,560	2,565	2,570	2,575	2,580	2,585	2,590
52	2,595	2,600	2,605	2,610	2,615	2,620	2,625	2,630	2,635	2,640
53	2,645	2,650	2,655	2,660	2,665	2,670	2,675	2,680	2,685	2,690
54	2,695	2,700	2,705	2,710	2,715	2,720	2,725	2,730	2,735	2,740
55	2,745	2,749	2,754	2,759	2,764	2,769	2,774	2,779	2,784	2,789
56	2,794	2,799	2,804	2,809	2,814	2,819	2,824	2,829	2,834	2,839
57	2,844	2,849	2,854	2,859	2,864	2,869	2,874	2,879	2,884	2,889
58	2,894	2,899	2,904	2,909	2,914	2,919	2,924	2,929	2,934	2,939
59	2,944	2,949	2,954	2,959	2,964	2,969	2,974	2,979	2,984	2,989
60	2,994	2,999	3,004	3,009	3,014	3,019	3,024	3,029	3,034	3,039
61	3,044	3,049	3,054	3,059	3,064	3,069	3,074	3,079	3,084	3,089
62	3,094	3,099	3,104	3,109	3,114	3,119	3,124	3,129	3,134	3,139
63	3,144	3,149	3,154	3,159	3,164	3,169	3,174	3,179	3,184	3,189
64	3,194	3,199	3,204	3,209	3,214	3,219	3,224	3,229	3,234	3,239
65	3,244	3,248	3,253	3,258	3,263	3,268	3,273	3,278	3,283	3,288
66	3,293	3,298	3,303	3,308	3,313	3,318	3,323	3,328	3,333	3,338
67	3,343	3,348	3,353	3,358	3,363	3,368	3,373	3,378	3,383	3,388
68	3,393	3,398	3,403	3,408	3,413	3,418	3,423	3,428	3,433	3,438
69	3,443	3,448	3,453	3,458	3,463	3,468	3,473	3,478	3,483	3,488
70	3,493	3,498	3,503	3,508	3,513	3,518	3,523	3,528	3,533	3,538
71	3,543	3,548	3,553	3,558	3,563	3,568	3,573	3,578	3,583	3,588
72	3,593	3,598	3,603	3,608	3,613	3,618	3,623	3,628	3,633	3,638
73	3,643	3,648	3,653	3,658	3,663	3,668	3,673	3,678	3,683	3,688
74	3,693	3,698	3,703	3,708	3,713	3,718	3,723	3,728	3,733	3,738
75	3,743	3,747	3,752	3,757	3,762	3,767	3,772	3,777	3,782	3,787
76	3,792	3,797	3,802	3,807	3,812	3,817	3,822	3,827	3,832	3,837
77	3,842	3,847	3,852	3,857	3,862	3,867	3,872	3,877	3,882	3,887
78	3,892	3,897	3,902	3,907	3,912	3,917	3,922	3,927	3,932	3,937
79	3,942	3,947	3,952	3,957	3,962	3,967	3,972	3,977	3,982	3,987
80	3,992	3,997	4,002	4,007	4,012	4,017	4,022	4,027	4,032	4,037
81	4,042	4,047	4,052	4,057	4,062	4,067	4,072	4,077	4,082	4,087
82	4,092	4,097	4,102	4,107	4,112	4,117	4,122	4,127	4,132	4,137
83	4,142	4,147	4,152	4,157	4,162	4,167	4,172	4,177	4,182	4,187
84	4,192	4,197	4,202	4,207	4,212	4,217	4,222	4,227	4,232	4,237
85	4,242	4,246	4,251	4,256	4,261	4,266	4,271	4,276	4,281	4,286
86	4,291	4,296	4,301	4,306	4,311	4,316	4,321	4,326	4,331	4,336
87	4,341	4,346	4,351	4,356	4,361	4,366	4,371	4,376	4,381	4,386
88	4,391	4,396	4,401	4,406	4,411	4,416	4,421	4,426	4,431	4,436
89	4,441	4,446	4,451	4,456	4,461	4,466	4,471	4,476	4,481	4,486
90	4,491	4,496	4,501	4,506	4,511	4,516	4,521	4,526	4,531	4,536
91	4,541	4,546	4,551	4,556	4,561	4,566	4,571	4,576	4,581	4,586
92	4,591	4,596	4,601	4,606	4,611	4,616	4,621	4,626	4,631	4,636
93	4,641	4,646	4,651	4,656	4,661	4,666	4,671	4,676	4,681	4,686
94	4,691	4,696	4,701	4,706	4,711	4,716	4,721	4,726	4,731	4,736
95	4,741	4,745	4,750	4,755	4,760	4,765	4,770	4,775	4,780	4,785
96	4,790	4,795	4,800	4,805	4,810	4,815	4,820	4,825	4,830	4,835
97	4,840	4,845	4,850	4,855	4,860	4,865	4,870	4,875	4,880	4,885
98	4,890	4,895	4,900	4,905	4,910	4,915	4,920	4,925	4,930	4,935
99	4,940	4,945	4,950	4,955	4,960	4,965	4,970	4,975	4,980	4,985
100	4,990	4,995	5,000	5,005	5,010	5,015	5,020	5,025	5,030	5,035

mg	1000	2000	3000	4000	5000	6000	7000	8000	9000	10000
meq	49,90	99,80	149,70	199,60	249,50	299,40	349,30	399,20	449,10	499,00

13

Table 5

Conversion of Milligrams of Mg^{2+} into Milligram-Equivalents
(equivalent weight of Mg^{2+} = 12.16)

Whole mg	Tenths of a mg									
	0	1	2	3	4	5	6	7	8	9
0	—	0,01	0,02	0,02	0,03	0,04	0,05	0,06	0,07	0,07
1	0,082	0,090	0,099	0,107	0,115	0,123	0,132	0,140	0,148	0,156
2	0,164	0,173	0,181	0,189	0,197	0,206	0,214	0,222	0,230	0,238
3	0,247	0,255	0,263	0,271	0,280	0,288	0,296	0,304	0,313	0,321
4	0,329	0,337	0,345	0,354	0,362	0,370	0,378	0,387	0,395	0,403
5	0,411	0,419	0,428	0,436	0,444	0,452	0,460	0,469	0,477	0,485
6	0,493	0,502	0,510	0,518	0,526	0,535	0,543	0,551	0,559	0,567
7	0,576	0,584	0,592	0,600	0,609	0,617	0,625	0,633	0,641	0,650
8	0,658	0,666	0,674	0,683	0,691	0,699	0,707	0,716	0,724	0,732
9	0,740	0,748	0,757	0,765	0,773	0,781	0,789	0,798	0,806	0,814
10	0,822	0,831	0,839	0,847	0,855	0,863	0,872	0,880	0,888	0,896
11	0,905	0,913	0,921	0,929	0,938	0,946	0,954	0,962	0,970	0,979
12	0,987	0,995	1,003	1,012	1,020	1,028	1,036	1,044	1,053	1,061
13	1,069	1,077	1,086	1,094	1,102	1,110	1,118	1,127	1,135	1,143
14	1,151	1,160	1,168	1,176	1,184	1,192	1,201	1,209	1,217	1,225
15	1,234	1,242	1,250	1,258	1,267	1,275	1,283	1,291	1,299	1,308
16	1,316	1,324	1,332	1,341	1,349	1,357	1,365	1,373	1,382	1,390
17	1,398	1,406	1,415	1,423	1,431	1,439	1,447	1,456	1,464	1,472
18	1,480	1,489	1,497	1,505	1,513	1,521	1,530	1,538	1,546	1,554
19	1,563	1,571	1,579	1,587	1,595	1,604	1,612	1,620	1,628	1,637
20	1,645	1,653	1,661	1,669	1,678	1,686	1,694	1,702	1,711	1,719
21	1,727	1,736	1,744	1,752	1,760	1,769	1,777	1,785	1,793	1,801
22	1,809	1,817	1,826	1,834	1,842	1,850	1,859	1,867	1,875	1,883
23	1,892	1,900	1,908	1,916	1,924	1,933	1,941	1,949	1,957	1,966
24	1,974	1,982	1,990	1,998	2,007	2,015	2,023	2,031	2,040	2,048
25	2,056	2,064	2,072	2,081	2,089	2,097	2,105	2,114	2,122	2,130
26	2,138	2,146	2,155	2,163	2,171	2,179	2,188	2,196	2,204	2,212
27	2,220	2,229	2,237	2,245	2,253	2,262	2,270	2,278	2,286	2,294
28	2,303	2,311	2,319	2,327	2,336	2,344	2,352	2,360	2,369	2,377
29	2,385	2,393	2,401	2,410	2,418	2,426	2,434	2,443	2,451	2,459
30	2,467	2,475	2,484	2,492	2,500	2,508	2,516	2,525	2,533	2,541
31	2,549	2,558	2,566	2,574	2,582	2,591	2,599	2,607	2,615	2,623
32	2,632	2,640	2,648	2,657	2,665	2,673	2,681	2,690	2,698	2,706
33	2,714	2,722	2,730	2,739	2,747	2,755	2,763	2,771	2,780	2,788
34	2,796	2,804	2,813	2,821	2,829	2,837	2,845	2,854	2,862	2,870
35	2,878	2,887	2,895	2,903	2,911	2,919	2,928	2,936	2,944	2,952
36	2,961	2,969	2,977	2,985	2,993	3,002	3,010	3,018	3,026	3,035
37	3,043	3,051	3,059	3,068	3,076	3,084	3,092	3,100	3,109	3,117
38	3,125	3,133	3,141	3,150	3,158	3,166	3,174	3,183	3,191	3,199
39	3,207	3,215	3,224	3,232	3,240	3,248	3,257	3,265	3,273	3,281
40	3,290	3,298	3,306	3,314	3,322	3,331	3,339	3,347	3,355	3,364
41	3,372	3,380	3,388	3,396	3,405	3,413	3,421	3,429	3,438	3,446
42	3,454	3,462	3,470	3,479	3,487	3,495	3,503	3,512	3,520	3,528
43	3,536	3,544	3,553	3,561	3,569	3,577	3,586	3,594	3,602	3,610
44	3,618	3,627	3,635	3,643	3,651	3,660	3,668	3,676	3,684	3,692
45	3,701	3,709	3,717	3,725	3,734	3,742	3,750	3,758	3,767	3,775
46	3,783	3,791	3,799	3,808	3,816	3,824	3,832	3,841	3,849	3,857
47	3,865	3,873	3,882	3,890	3,898	3,906	3,915	3,923	3,931	3,939
48	3,947	3,956	3,964	3,972	3,980	3,989	3,997	4,005	4,013	4,021
49	4,030	4,038	4,046	4,054	4,063	4,071	4,079	4,087	4,095	4,104
50	4,112	4,120	4,128	4,137	4,145	4,153	4,161	4,170	4,178	4,186

Whole mg	Tenths of a mg									
	0	1	2	3	4	5	6	7	8	9
51	4,194	4,202	4,211	4,219	4,227	4,235	4,243	4,252	4,260	4,268
52	4,276	4,285	4,293	4,301	4,309	4,317	4,326	4,334	4,342	4,350
53	4,359	4,367	4,375	4,383	4,392	4,400	4,408	4,416	4,424	4,433
54	4,441	4,449	4,457	4,466	4,474	4,482	4,490	4,498	4,507	4,515
55	4,523	4,531	4,539	4,548	4,556	4,564	4,572	4,581	4,589	4,597
56	4,605	4,614	4,622	4,630	4,638	4,646	4,655	4,663	4,671	4,679
57	4,688	4,696	4,704	4,712	4,720	4,729	4,737	4,745	4,753	4,762
58	4,770	4,778	4,786	4,795	4,803	4,811	4,819	4,827	4,836	4,844
59	4,852	4,860	4,868	4,877	4,885	4,893	4,901	4,910	4,918	4,926
60	4,934	4,942	4,951	4,959	4,967	4,975	4,984	4,992	5,000	5,009
61	5,017	5,025	5,033	5,041	5,049	5,058	5,066	5,074	5,082	5,091
62	5,099	5,107	5,115	5,123	5,132	5,140	5,148	5,156	5,165	5,173
63	5,181	5,189	5,197	5,206	5,214	5,222	5,230	5,239	5,247	5,255
64	5,263	5,271	5,280	5,288	5,296	5,304	5,313	5,321	5,329	5,337
65	5,345	5,354	5,362	5,370	5,378	5,387	5,395	5,403	5,411	5,419
66	5,428	5,436	5,444	5,452	5,461	5,469	5,477	5,485	5,493	5,502
67	5,510	5,518	5,526	5,535	5,543	5,551	5,559	5,568	5,576	5,584
68	5,592	5,600	5,609	5,617	5,625	5,633	5,641	5,650	5,658	5,666
69	5,674	5,683	5,691	5,699	5,707	5,716	5,724	5,732	5,740	5,748
70	5,757	5,765	5,773	5,781	5,790	5,798	5,806	5,814	5,822	5,831
71	5,839	5,847	5,855	5,864	5,872	5,880	5,888	5,897	5,905	5,913
72	5,921	5,929	5,937	5,946	5,954	5,962	5,970	5,979	5,987	5,995
73	6,003	6,012	6,020	6,028	6,036	6,044	6,053	6,061	6,069	6,077
74	6,086	6,094	6,102	6,110	6,119	6,127	6,135	6,143	6,151	6,160
75	6,168	6,176	6,184	6,193	6,201	6,209	6,217	6,226	6,234	6,242
76	6,250	6,258	6,266	6,275	6,283	6,291	6,299	6,308	6,316	6,324
77	6,332	6,341	6,349	6,357	6,365	6,373	6,382	6,390	6,398	6,406
78	6,414	6,423	6,431	6,439	6,447	6,456	6,464	6,472	6,480	6,488
79	6,497	6,505	6,513	6,522	6,530	6,538	6,545	6,554	6,563	6,571
80	6,580	6,587	6,595	6,604	6,612	6,620	6,628	6,637	6,645	6,653
81	6,661	6,670	6,678	6,686	6,694	6,702	6,711	6,719	6,727	6,735
82	6,743	6,752	6,760	6,768	6,776	6,785	6,793	6,801	6,809	6,817
83	6,826	6,834	6,842	6,850	6,859	6,867	6,875	6,883	6,892	6,900
84	6,908	6,916	6,924	6,933	6,941	6,949	6,957	6,966	6,974	6,982
85	6,990	6,998	7,007	7,015	7,023	7,031	7,040	7,048	7,056	7,064
86	7,073	7,081	7,089	7,097	7,105	7,114	7,122	7,130	7,138	7,147
87	7,155	7,163	7,171	7,179	7,188	7,196	7,204	7,212	7,221	7,229
88	7,237	7,245	7,253	7,262	7,270	7,278	7,286	7,294	7,303	7,311
89	7,319	7,327	7,336	7,344	7,352	7,360	7,369	7,377	7,385	7,393
90	7,401	7,410	7,418	7,426	7,434	7,442	7,451	7,459	7,467	7,475
91	7,484	7,492	7,500	7,508	7,517	7,525	7,533	7,541	7,549	7,558
92	7,566	7,574	7,582	7,591	7,599	7,607	7,615	7,623	7,632	7,640
93	7,648	7,656	7,664	7,673	7,681	7,689	7,697	7,706	7,714	7,722
94	7,730	7,739	7,747	7,755	7,763	7,771	7,780	7,788	7,796	7,804
95	7,813	7,821	7,829	7,837	7,845	7,854	7,862	7,870	7,878	7,887
96	7,895	7,903	7,911	7,920	7,928	7,936	7,944	7,952	7,961	7,969
97	7,977	7,985	7,993	8,002	8,010	8,018	8,026	8,035	8,043	8,051
98	8,059	8,068	8,076	8,084	8,092	8,100	8,109	8,117	8,125	8,133
99	8,142	8,150	8,158	8,166	8,175	8,183	8,191	8,199	8,207	8,216
100	8,224	8,232	8,240	8,249	8,257	8,265	8,273	8,281	8,290	8,298

mg	1000	2000	3000	4000	5000	6000	7000	8000	9000	10 000
meq	82,24	164,47	246,71	328,95	411,18	493,42	575,66	657,90	740,13	822,37

Table 6

Conversion of Milligrams of K$^+$ into Milligram-Equivalents
(equivalent weight of K$^+$ = 39.100)

Whole mg	Tenths of a mg									
	0	1	2	3	4	5	6	7	8	9
0	—	0,00	0,01	0,01	0,01	0,01	0,02	0,02	0,02	0,02
1	0,026	0,028	0,031	0,033	0,036	0,038	0,041	0,043	0,046	0,049
2	0,051	0,054	0,056	0,059	0,061	0,064	0,067	0,069	0,072	0,074
3	0,077	0,079	0,082	0,084	0,087	0,090	0,092	0,095	0,097	0,100
4	0,102	0,105	0,107	0,110	0,113	0,115	0,118	0,120	0,123	0,125
5	0,128	0,130	0,133	0,136	0,138	0,141	0,143	0,146	0,148	0,151
6	0,153	0,156	0,159	0,161	0,164	0,166	0,169	0,171	0,174	0,176
7	0,179	0,182	0,184	0,187	0,189	0,192	0,194	0,197	0,200	0,202
8	0,205	0,207	0,210	0,212	0,215	0,217	0,220	0,223	0,225	0,228
9	0,230	0,233	0,235	0,238	0,240	0,243	0,246	0,248	0,251	0,253
10	0,256	0,258	0,261	0,263	0,266	0,269	0,271	0,274	0,276	0,279
11	0,281	0,284	0,286	0,289	0,292	0,294	0,297	0,299	0,302	0,304
12	0,307	0,309	0,312	0,315	0,317	0,320	0,322	0,325	0,327	0,330
13	0,332	0,335	0,338	0,340	0,343	0,345	0,348	0,350	0,353	0,356
14	0,358	0,361	0,363	0,366	0,368	0,371	0,373	0,376	0,379	0,381
15	0,384	0,386	0,389	0,391	0,394	0,396	0,399	0,402	0,404	0,407
16	0,409	0,412	0,414	0,417	0,419	0,422	0,425	0,427	0,430	0,432
17	0,435	0,437	0,440	0,442	0,445	0,448	0,450	0,453	0,455	0,458
18	0,460	0,463	0,465	0,468	0,471	0,473	0,476	0,478	0,481	0,483
19	0,486	0,488	0,491	0,494	0,496	0,499	0,501	0,504	0,506	0,509
20	0,512	0,514	0,517	0,519	0,522	0,524	0,527	0,529	0,532	0,535
21	0,537	0,540	0,542	0,545	0,547	0,550	0,552	0,555	0,558	0,560
22	0,563	0,565	0,568	0,570	0,573	0,575	0,578	0,581	0,583	0,586
23	0,588	0,591	0,593	0,596	0,598	0,601	0,604	0,606	0,609	0,611
24	0,614	0,616	0,619	0,621	0,624	0,627	0,629	0,632	0,634	0,637
25	0,639	0,642	0,644	0,647	0,650	0,652	0,655	0,657	0,660	0,662
26	0,665	0,668	0,670	0,673	0,675	0,678	0,680	0,683	0,685	0,688
27	0,691	0,693	0,696	0,698	0,701	0,703	0,706	0,708	0,711	0,714
28	0,716	0,719	0,721	0,724	0,726	0,729	0,731	0,734	0,737	0,739
29	0,742	0,744	0,747	0,749	0,752	0,754	0,757	0,760	0,762	0,765
30	0,767	0,770	0,772	0,775	0,777	0,780	0,783	0,785	0,788	0,790
31	0,793	0,795	0,798	0,800	0,803	0,806	0,808	0,811	0,813	0,816
32	0,818	0,821	0,824	0,826	0,829	0,831	0,834	0,836	0,839	0,841
33	0,844	0,847	0,849	0,852	0,854	0,857	0,859	0,862	0,864	0,867
34	0,870	0,872	0,875	0,877	0,880	0,882	0,885	0,887	0,890	0,893
35	0,895	0,898	0,900	0,903	0,905	0,908	0,910	0,913	0,916	0,918
36	0,921	0,923	0,926	0,928	0,931	0,934	0,936	0,939	0,941	0,944
37	0,946	0,949	0,951	0,954	0,957	0,959	0,962	0,964	0,967	0,969
38	0,972	0,974	0,977	0,980	0,982	0,985	0,987	0,990	0,992	0,995
39	0,997	1,000	1,003	1,005	1,008	1,010	1,013	1,015	1,018	1,020
40	1,023	1,026	1,028	1,031	1,033	1,036	1,038	1,041	1,043	1,046
41	1,049	1,051	1,054	1,056	1,059	1,061	1,064	1,066	1,069	1,072
42	1,074	1,077	1,079	1,082	1,084	1,087	1,090	1,092	1,095	1,097
43	1,100	1,102	1,105	1,107	1,110	1,113	1,115	1,118	1,120	1,123
44	1,125	1,128	1,130	1,133	1,136	1,138	1,141	1,143	1,146	1,148
45	1,151	1,153	1,156	1,159	1,161	1,164	1,166	1,169	1,171	1,174
46	1,176	1,179	1,182	1,184	1,187	1,189	1,192	1,194	1,197	1,199
47	1,202	1,205	1,207	1,210	1,212	1,215	1,217	1,220	1,223	1,225
48	1,228	1,230	1,233	1,235	1,238	1,240	1,243	1,246	1,248	1,251
49	1,253	1,256	1,258	1,261	1,263	1,266	1,269	1,271	1,274	1,276
50	1,279	1,281	1,284	1,286	1,289	1,292	1,294	1,297	1,299	1,302

Table 7

Conversion of Milligrams of Fe^{2+} into Milligram-Equivalents
(equivalent weight of Fe^{2+} = 27.925)

Whole mg	Tenths of a mg									
	0	1	2	3	4	5	6	7	8	9
0	—	0,00	0,01	0,01	0,01	0,02	0,02	0,03	0,03	0,03
1	0,036	0,039	0,043	0,047	0,050	0,054	0,057	0,061	0,065	0,068
2	0,072	0,075	0,079	0,082	0,086	0,090	0,093	0,097	0,100	0,104
3	0,107	0,110	0,115	0,118	0,122	0,125	0,129	0,133	0,136	0,140
4	0,143	0,147	0,150	0,154	0,158	0,161	0,165	0,168	0,172	0,175
5	0,179	0,183	0,186	0,190	0,193	0,197	0,201	0,204	0,208	0,211
6	0,215	0,218	0,222	0,226	0,229	0,233	0,236	0,240	0,244	0,247
7	0,251	0,254	0,258	0,261	0,265	0,269	0,272	0,276	0,279	0,283
8	0,286	0,290	0,294	0,297	0,301	0,304	0,308	0,312	0,315	0,319
9	0,322	0,326	0,329	0,333	0,337	0,340	0,344	0,347	0,351	0,355
10	0,358	0,362	0,365	0,369	0,372	0,376	0,380	0,383	0,387	0,390
11	0,394	0,397	0,401	0,405	0,408	0,412	0,415	0,419	0,423	0,426
12	0,430	0,433	0,437	0,440	0,444	0,448	0,451	0,455	0,458	0,462
13	0,466	0,469	0,473	0,476	0,480	0,483	0,487	0,491	0,494	0,498
14	0,501	0,505	0,509	0,512	0,516	0,519	0,523	0,526	0,530	0,534
15	0,537	0,541	0,544	0,548	0,551	0,555	0,559	0,562	0,566	0,569
16	0,573	0,577	0,580	0,584	0,587	0,591	0,594	0,598	0,602	0,605
17	0,609	0,612	0,616	0,620	0,623	0,627	0,630	0,634	0,637	0,641
18	0,645	0,648	0,652	0,655	0,659	0,662	0,666	0,670	0,673	0,677
19	0,680	0,684	0,688	0,691	0,695	0,698	0,702	0,705	0,709	0,713
20	0,716	0,720	0,723	0,727	0,731	0,734	0,738	0,741	0,745	0,748
21	0,752	0,756	0,759	0,763	0,766	0,770	0,773	0,777	0,781	0,784
22	0,788	0,791	0,795	0,799	0,802	0,806	0,809	0,813	0,816	0,820
23	0,824	0,827	0,831	0,834	0,838	0,842	0,845	0,849	0,852	0,856
24	0,859	0,863	0,867	0,870	0,874	0,877	0,881	0,885	0,888	0,892
25	0,895	0,899	0,902	0,906	0,910	0,913	0,917	0,920	0,924	0,927
26	0,931	0,935	0,938	0,942	0,945	0,949	0,953	0,956	0,960	0,963
27	0,967	0,970	0,974	0,978	0,981	0,985	0,988	0,992	0,996	0,999
28	1,003	1,006	1,010	1,013	1,017	1,021	1,024	1,028	1,031	1,035
29	1,038	1,042	1,046	1,049	1,053	1,056	1,060	1,064	1,067	1,071
30	1,074	1,078	1,081	1,085	1,089	1,092	1,096	1,099	1,103	1,107
31	1,110	1,114	1,117	1,121	1,124	1,128	1,132	1,135	1,139	1,142
32	1,146	1,150	1,153	1,157	1,160	1,164	1,167	1,171	1,175	1,178
33	1,182	1,185	1,189	1,192	1,196	1,200	1,203	1,207	1,210	1,214
34	1,218	1,221	1,225	1,228	1,232	1,235	1,239	1,243	1,246	1,250
35	1,253	1,257	1,261	1,264	1,268	1,271	1,275	1,278	1,282	1,286
36	1,289	1,293	1,296	1,300	1,303	1,307	1,311	1,314	1,318	1,321
37	1,325	1,329	1,332	1,336	1,339	1,343	1,346	1,350	1,354	1,357
38	1,361	1,364	1,368	1,372	1,375	1,379	1,382	1,386	1,389	1,393
39	1,397	1,400	1,404	1,407	1,411	1,414	1,418	1,422	1,425	1,429
40	1,432	1,436	1,440	1,443	1,447	1,450	1,454	1,457	1,461	1,465
41	1,468	1,472	1,476	1,479	1,483	1,486	1,490	1,493	1,497	1,500
42	1,504	1,508	1,511	1,515	1,518	1,522	1,526	1,529	1,533	1,536
43	1,540	1,543	1,547	1,551	1,554	1,558	1,561	1,565	1,568	1,572
44	1,576	1,579	1,583	1,586	1,590	1,594	1,597	1,601	1,604	1,608
45	1,611	1,615	1,619	1,622	1,626	1,629	1,633	1,637	1,640	1,644
46	1,647	1,651	1,654	1,658	1,662	1,665	1,669	1,672	1,676	1,679
47	1,683	1,687	1,690	1,694	1,697	1,701	1,705	1,708	1,712	1,715
48	1,719	1,722	1,726	1,730	1,733	1,737	1,740	1,744	1,748	1,751
49	1,755	1,758	1,762	1,765	1,769	1,773	1,776	1,780	1,783	1,787
50	1,791	1,794	1,798	1,801	1,805	1,808	1,812	1,816	1,819	1,823

Table 8

Conversion of Milligrams of Fe^{3+} into Milligram-Equivalents
(equivalent weight of Fe^{3+} = 18.617)

Whole mg	Tenths of a mg									
	0	1	2	3	4	5	6	7	8	9
0	—	0,01	0,01	0,02	0,02	0,03	0,03	0,04	0,04	0,05
1	0,054	0,059	0,064	0,070	0,075	0,081	0,086	0,091	0,097	0,102
2	0,107	0,113	0,118	0,124	0,129	0,134	0,140	0,145	0,150	0,156
3	0,161	0,167	0,172	0,177	0,182	0,188	0,193	0,199	0,204	0,210
4	0,215	0,220	0,226	0,231	0,236	0,242	0,247	0,253	0,258	0,263
5	0,269	0,274	0,279	0,285	0,290	0,295	0,301	0,306	0,312	0,317
6	0,322	0,328	0,333	0,338	0,344	0,349	0,355	0 360	0,365	0,371
7	0,376	0,381	0,387	0,392	0,398	0,403	0,408	0,414	0,419	0,424
8	0,430	0,435	0,441	0,446	0,451	0,457	0,462	0,467	0,473	0,478
9	0,484	0,489	0,494	0,500	0,505	0,510	0,516	0,521	0,527	0,532
10	0,537	0,543	0,548	0,553	0,559	0,564	0,569	0,575	0,580	0,586

Table 9

Conversion of Milligrams of Al^{3+} into Milligram-Equivalents
(equivalent weight of Al^{3+} = 8.993)

Whole mg	Tenths of a mg									
	0	1	2	3	4	5	6	7	8	9
0	—	0,01	0,02	0,03	0,04	0,06	0,07	0,08	0,09	0,10
1	0,111	0,122	0,133	0,145	0,156	0,167	0,178	0,189	0,200	0,211
2	0,222	0,234	0,245	0,256	0,267	0,278	0,289	0,300	0,311	0,323
3	0 334	0,345	0,356	0,367	0,378	0,389	0,400	0,412	0,423	0,434
4	0,445	0,456	0,467	0,478	0,489	0,501	0,512	0,523	0,534	0,545
5	0,556	0,567	0,578	0,590	0,601	0,612	0,623	0,634	0,645	0,656
6	0,667	0,679	0,690	0,701	0,712	0,723	0,734	0,745	0,756	0,767
7	0,779	0,790	0,801	0,812	0,823	0,834	0,845	0,856	0,868	0,879
8	0,890	0,901	0,912	0,923	0,934	0,945	0,957	0,968	0,979	0,990
9	1,001	1,012	1,023	1,034	1,046	1,057	1,068	1,079	1,090	1,101
10	1,112	1,123	1,135	1,146	1,157	1,168	1,179	1,190	1,201	1,212

Table 10

Conversion of Milligrams of Mn^{2+} into Milligram-Equivalents
(equivalent weight of Mn^{2+} = 27.47)

Whole mg	Tenths of a mg									
	0	1	2	3	4	5	6	7	8	9
0	—	0,00	0,01	0,01	0,01	0 02	0,02	0,03	0,03	0,03
1	0,036	0,040	0,044	0,047	0,051	0,055	0,058	0,062	0,066	0,069
2	0,073	0,076	0,080	0,084	0,087	0,091	0,095	0,098	0,102	0,106
3	0,109	0,113	0,117	0,120	0,124	0,127	0,131	0,135	0,138	0,142
4	0,146	0,149	0,153	0,157	0,160	0,164	0,167	0,171	0,175	0,178
5	0,182	0,186	0,189	0,193	0,197	0,200	0,204	0,208	0,211	0,215
6	0,218	0,222	0,226	0,229	0,233	0,237	0,240	0,244	0,248	0,251
7	0,255	0,259	0,262	0,266	0,269	0,273	0,277	0,280	0,284	0,288
8	0,291	0,295	0,299	0,302	0,306	0,309	0,313	0,317	0,320	0,324
9	0,328	0,331	0,335	0,339	0,342	0,346	0,350	0,353	0,357	0,360
10	0,364	0,368	0,371	0,375	0,379	0,382	0,386	0,390	0,393	0,397

Table 11

Conversion of Milligrams of NH_4^+ into Milligram-Equivalents
(equivalent weight of NH_4^+ = 18.040)

Whole mg	Tenths of a mg									
	0	1	2	3	4	5	6	7	8	9
0	—	0,01	0,01	0,02	0,02	0,03	0,03	0,04	0,04	0,05
1	0,055	0,061	0,067	0,072	0,078	0,083	0,089	0,094	0,100	0,105
2	0,111	0,116	0,122	0,128	0,133	0,139	0,144	0,150	0,155	0,161
3	0,166	0,172	0,177	0,183	0,188	0,194	0,200	0,205	0,211	0,216
4	0,222	0,227	0,233	0,238	0,244	0,249	0,255	0,261	0,266	0,272
5	0,277	0,283	0,288	0,294	0,299	0,305	0,310	0,316	0,322	0,327
6	0,333	0,338	0,344	0,349	0,355	0,360	0,366	0,371	0,377	0,383
7	0,388	0,394	0,399	0,405	0,410	0,416	0,421	0,427	0,432	0,438
8	0,443	0,449	0,455	0,460	0,466	0,471	0,477	0 482	0,488	0,493
9	0,499	0,504	0,510	0,516	0,521	0,527	0,532	0,538	0,543	0,549
10	0,554	0,560	0,565	0,571	0,577	0,582	0,588	0,593	0,599	0,604

Table 12

Conversion of Milligrams of Cl⁻ into Milligram—Equivalents
(equivalent weight of $Cl^- = 35.457$)

Whole mg	Tenths of a mg									
	0	1	2	3	4	5	6	7	8	9
0	—	0,00	0,01	0,01	0,01	0,01	0,02	0,02	0,02	0,03
1	0,028	0,031	0,034	0,037	0,039	0,042	0,045	0,048	0,051	0,054
2	0,056	0,059	0,062	0,065	0,068	0,071	0,073	0,076	0,079	0,082
3	0,085	0,087	0,090	0,093	0,096	0,099	0,102	0,104	0,107	0,110
4	0,113	0,116	0,119	0,121	0,124	0,127	0,130	0,133	0,135	0,138
5	0,141	0,144	0,147	0,150	0,152	0,155	0,158	0,161	0,164	0,166
6	0,169	0,172	0,175	0,178	0,181	0,183	0,186	0,189	0,192	0,195
7	0,197	0,200	0,203	0,206	0,209	0,212	0,214	0,217	0,220	0,223
8	0,226	0,228	0,231	0,234	0,237	0,240	0,243	0,245	0,248	0,251
9	0,254	0,257	0,260	0,262	0,265	0,268	0,271	0,274	0,276	0,279
10	0,282	0,285	0,288	0,291	0,293	0,296	0,299	0,302	0,305	0,307
11	0,310	0,313	0,316	0,319	0,322	0,324	0,327	0,330	0,333	0,336
12	0,338	0,341	0,344	0,347	0,350	0,353	0,355	0,358	0,361	0,364
13	0,367	0,369	0,372	0,375	0,378	0,381	0,384	0,386	0,389	0,392
14	0,395	0,398	0,401	0,403	0,406	0,409	0,412	0,415	0,417	0,420
15	0,423	0,426	0,429	0,432	0,434	0,437	0,440	0,443	0,446	0,449
16	0,451	0,454	0,457	0,460	0,463	0,465	0,468	0,471	0,474	0,477
17	0,480	0,482	0,485	0,488	0,491	0,494	0,496	0,499	0,502	0,505
18	0,508	0,511	0,513	0,516	0,519	0,522	0,525	0,527	0,530	0,533
19	0,536	0,539	0,542	0,544	0,547	0,550	0,553	0,556	0,559	0,561
20	0,564	0,567	0,570	0,573	0,575	0,578	0,581	0,584	0,587	0,590
21	0,592	0,595	0,598	0,601	0,604	0,606	0,609	0,612	0,615	0,618
22	0,621	0,623	0,626	0,629	0,632	0,635	0,637	0,640	0,643	0,646
23	0,649	0,652	0,654	0,657	0,660	0,663	0,666	0,668	0,671	0,674
24	0,677	0,680	0,683	0,685	0,688	0,691	0,694	0,697	0,700	0,702
25	0,705	0,708	0,711	0,714	0,716	0,719	0,722	0,725	0,728	0,731
26	0,733	0,736	0,739	0,742	0,745	0,747	0,750	0,753	0,756	0,759
27	0,762	0,764	0,767	0,770	0,773	0,776	0,778	0,781	0,784	0,787
28	0,790	0,793	0,795	0,798	0,801	0,804	0,807	0,809	0,812	0,815
29	0,818	0,821	0,824	0,826	0,829	0,832	0,835	0,838	0,841	0,843
30	0,846	0,849	0,852	0,855	0,857	0,860	0,863	0,866	0,869	0,872
31	0,874	0,877	0,880	0,883	0,886	0,888	0,891	0,894	0,897	0,900
32	0,903	0,905	0,908	0,911	0,914	0,917	0,919	0,922	0,925	0,928
33	0,931	0,934	0,936	0,939	0,942	0,945	0,948	0,950	0,953	0,956
34	0,959	0,962	0,965	0,967	0,970	0,973	0,976	0,979	0,982	0,984
35	0,987	0,990	0,993	0,996	0,998	1,001	1,004	1,007	1,010	1,013
36	1,015	1,018	1,021	1,024	1,027	1,029	1,032	1,035	1,038	1,041
37	1,044	1,046	1,049	1,052	1,055	1,058	1,060	1,063	1,066	1,069
38	1,072	1,075	1,077	1,080	1,083	1,086	1,089	1,091	1,094	1,097
39	1,100	1,103	1,106	1,108	1,111	1,114	1,117	1,120	1,123	1,125
40	1,128	1,131	1,134	1,137	1,139	1,142	1,145	1,148	1,151	1,154
41	1,156	1,159	1,162	1,165	1,168	1,170	1,173	1,176	1,179	1,182
42	1,185	1,187	1,190	1,193	1,196	1,199	1,201	1,204	1,207	1,210
43	1,213	1,216	1,218	1,221	1,224	1,227	1,230	1,233	1,235	1,238
44	1,241	1,244	1,247	1,249	1,252	1,255	1,259	1,261	1,264	1,266
45	1,269	1,272	1,275	1,278	1,280	1,283	1,286	1,289	1,292	1,295
46	1,297	1,300	1,303	1,306	1,309	1,312	1,314	1,317	1,320	1,323
47	1,326	1,328	1,331	1,334	1,337	1,340	1,343	1,345	1,348	1,351
48	1,354	1,357	1,359	1,362	1,365	1,368	1,371	1,374	1,376	1,379
49	1,382	1,385	1,388	1,390	1,393	1,396	1,399	1,402	1,405	1,407
50	1,410	1,413	1,416	1,419	1,422	1,424	1,427	1,430	1,433	1,436

Whole mg	Tenths of a mg									
	0	1	2	3	4	5	6	7	8	9
51	1,438	1,441	1,444	1,447	1,450	1,453	1,455	1,458	1,461	1,464
52	1,467	1,469	1,472	1,475	1,478	1,481	1,484	1,486	1,489	1,492
53	1,495	1,498	1,500	1,503	1,506	1,509	1,512	1,515	1,517	1,520
54	1,523	1,526	1,529	1,531	1,534	1,537	1,540	1,543	1,546	1,548
55	1,551	1,554	1,557	1,560	1,563	1,565	1,568	1,571	1,574	1,577
56	1,579	1,582	1,585	1,588	1,591	1,594	1,596	1,599	1,602	1,605
57	1,608	1,610	1,613	1,616	1,619	1,622	1,625	1,627	1,630	1,633
58	1,636	1,639	1,641	1,644	1,647	1,650	1,653	1,656	1,658	1,661
59	1,664	1,667	1,670	1,673	1,675	1,678	1,681	1,684	1,687	1,689
60	1,692	1,695	1,698	1,701	1,704	1,706	1,709	1,712	1,715	1,718
61	1,720	1,723	1,726	1,729	1,732	1,735	1,737	1,740	1,743	1,746
62	1,749	1,751	1,754	1,757	1,760	1,763	1,766	1,768	1,771	1,774
63	1,777	1,780	1,782	1,785	1,788	1,791	1,794	1,797	1,799	1,802
64	1,805	1,808	1,811	1,814	1,816	1,819	1,822	1,825	1,828	1,830
65	1,833	1,836	1,839	1,842	1,845	1,847	1,850	1,853	1,856	1,859
66	1,861	1,864	1,867	1,870	1,873	1,876	1,878	1,881	1,884	1,887
67	1,890	1,892	1,895	1,898	1,901	1,904	1,907	1,909	1,912	1,915
68	1,918	1,921	1,923	1,926	1,929	1,932	1,935	1,938	1,940	1,943
69	1,946	1,949	1,952	1,955	1,957	1,960	1,963	1,966	1,969	1,972
70	1,974	1,979	1,980	1,983	1,986	1,988	1,991	1,994	1,997	2,000
71	2,002	2,005	2,008	2,011	2,014	2,017	2,019	2,022	2,025	2,028
72	2,031	2,033	2,036	2,039	2,042	2,045	2,048	2,050	2,053	2,056
73	2,059	2,062	2,064	2,067	2,070	2,073	2,076	2,079	2,081	2,084
74	2,087	2,090	2,093	2,096	2,098	2,101	2,104	2,107	2,110	2,112
75	2,115	2,118	2,121	2,124	2,127	2,129	2,132	2,135	2,138	2,141
76	2,143	2,146	2,149	2,152	2,155	2,158	2,160	2,163	2,166	2,169
77	2,172	2,175	2,177	2,180	2,183	2,186	2,189	2,191	2,194	2,197
78	2,200	2,203	2,205	2,208	2,211	2,214	2,217	2,220	2,222	2,225
79	2,228	2,231	2,234	2,237	2,239	2,242	2,245	2,248	2,251	2,254
80	2,256	2,259	2,262	2,265	2,268	2,270	2,273	2,276	2,279	2,282
81	2,284	2,287	2,290	2,293	2,296	2,299	2,301	2,304	2,307	2,310
82	2,313	2,315	2,318	2,321	2,324	2,327	2,330	2,332	2,335	2,338
83	2,341	2,344	2,347	2,349	2,352	2,355	2,358	2,361	2,364	2,366
84	2,369	2,372	2,375	2,378	2,380	2,383	2,386	2,389	2,392	2,395
85	2,397	2,400	2,403	2,406	2,409	2,411	2,414	2,417	2,420	2,423
86	2,426	2,428	2,431	2,434	2,437	2,440	2,442	2,445	2,448	2,451
87	2,454	2,457	2,459	2,462	2,465	2,468	2,471	2,473	2,476	2,479
88	2,482	2,485	2,488	2,490	2,493	2,496	2,499	2,502	2,505	2,507
89	2,510	2,513	2,516	2,519	2,521	2,524	2,527	2,530	2,533	2,536
90	2,538	2,541	2,544	2,547	2,550	2,552	2,555	2,558	2,561	2,564
91	2,566	2,569	2,572	2,575	2,578	2,581	2,583	2,586	2,589	2,592
92	2,595	2,598	2,600	2,603	2,606	2,609	2,612	2,614	2,617	2,620
93	2,623	2,626	2,629	2,631	2,634	2,637	2,640	2,643	2,646	2,648
94	2,651	2,654	2,657	2,660	2,662	2,665	2,668	2,671	2,674	2,677
95	2,679	2,682	2,685	2,688	2,691	2,693	2,696	2,699	2,702	2,705
96	2,708	2,710	2,713	2,716	2,719	2,722	2,724	2,727	2,730	2,733
97	2,736	2,739	2,741	2,744	2,747	2,750	2,753	2,755	2,758	2,761
98	2,764	2,767	2,770	2,772	2,775	2,778	2,781	2,784	2,787	2,789
99	2,792	2,795	2,798	2,801	2,803	2,806	2,809	2,812	2,815	2,818
100	2,820	2,823	2,826	2,829	2,831	2,834	2,837	2,840	2,843	2,845

mg	1000	2000	3000	4000	5000	6000	7000	8000	9000	10 000
meq	28,20	56,41	84,61	112,81	141,02	169,22	197,42	225,62	253,83	282,03

Table 13

Conversion of Milligrams of SO_4^{2-} into Milligram—Equivalents
(equivalent weight of SO_4^{2-} = 48.033)

Whole mg	Tenths of a mg									
	0	1	2	3	4	5	6	7	8	9
0	—	0,00	0,00	0,01	0,01	0,01	0,01	0,01	0,02	0,02
1	0,021	0,023	0,025	0,027	0,029	0,031	0,033	0,035	0,037	0,040
2	0,042	0,044	0,046	0,048	0,050	0,052	0,054	0,056	0,058	0,060
3	0,062	0,065	0,067	0,069	0,071	0,073	0,075	0,077	0,079	0,081
4	0,083	0,085	0,087	0,090	0,092	0,094	0,096	0,098	0,100	0,102
5	0,104	0,106	0,108	0,110	0,112	0,115	0,117	0,119	0,121	0,123
6	0,125	0,127	0,129	0,131	0,133	0,135	0,137	0,139	0,142	0,144
7	0,146	0,148	0,150	0,152	0,154	0,156	0,158	0,160	0,162	0,164
8	0,167	0,169	0,171	0,173	0,175	0,177	0,179	0,181	0,183	0,185
9	0,187	0,189	0,192	0,194	0,196	0,198	0,200	0,202	0,204	0,206
10	0,208	0,210	0,212	0,214	0,217	0,219	0,221	0,223	0,225	0,227
11	0,229	0,231	0,233	0,235	0,237	0,239	0,242	0,244	0,246	0,248
12	0,250	0,252	0,254	0,256	0,258	0,260	0,262	0,264	0,266	0,269
13	0,271	0,273	0,275	0,277	0,279	0,281	0,283	0,285	0,287	0,289
14	0,291	0,294	0,296	0,298	0,300	0,302	0,304	0,306	0,308	0,310
15	0,312	0,314	0,316	0,319	0,321	0,323	0,325	0,327	0,329	0,331
16	0,333	0,335	0,337	0,339	0,341	0,344	0,346	0,348	0,350	0,352
17	0,354	0,356	0,358	0,360	0,362	0,364	0,366	0,369	0,371	0,373
18	0,375	0,377	0,379	0,381	0,383	0,385	0,387	0,389	0,391	0,393
19	0,396	0,398	0,400	0,402	0,404	0,406	0,408	0,410	0,412	0,414
20	0,416	0,418	0,421	0,423	0,425	0,427	0,429	0,431	0,433	0,435
21	0,437	0,439	0,441	0,443	0,446	0,448	0,450	0,452	0,454	0,456
22	0,458	0,460	0,462	0,464	0,466	0,468	0,471	0,473	0,475	0,477
23	0,479	0,481	0,483	0,485	0,487	0,489	0,491	0,493	0,496	0,498
24	0,500	0,502	0,504	0,506	0,508	0,510	0,512	0,514	0,516	0,518
25	0,520	0,523	0,525	0,527	0,529	0,531	0,533	0,535	0,537	0,539
26	0,541	0,543	0,545	0,548	0,550	0,552	0,554	0,556	0,558	0,560
27	0,562	0,564	0,566	0,568	0,570	0,573	0,575	0,577	0,579	0,581
28	0,583	0,585	0,587	0,589	0,591	0,593	0,595	0,598	0,600	0,602
29	0,604	0,606	0,608	0,610	0,612	0,614	0,616	0,618	0,620	0,622
30	0,625	0,627	0,629	0,631	0,633	0,635	0,637	0,639	0,641	0,643
31	0,645	0,647	0,650	0,652	0,654	0,656	0,658	0,660	0,662	0,664
32	0,666	0,668	0,670	0,672	0,675	0,677	0,679	0,681	0,683	0,685
33	0,687	0,689	0,691	0,693	0,695	0,697	0,700	0,702	0,704	0,706
34	0,708	0,710	0,712	0,714	0,716	0,718	0,720	0,722	0,725	0,727
35	0,729	0,731	0,733	0,735	0,737	0,739	0,741	0,743	0,745	0,747
36	0,749	0,752	0,754	0,756	0,758	0,760	0,762	0,764	0,766	0,768
37	0,770	0,772	0,774	0,777	0,779	0,781	0,783	0,785	0,787	0,789
38	0,791	0,793	0,795	0,797	0,799	0,802	0,804	0,806	0,808	0,810
39	0,812	0,814	0,816	0,818	0,820	0,822	0,824	0,827	0,829	0,831
40	0,833	0,835	0,837	0,839	0,841	0,843	0,845	0,847	0,849	0,851
41	0,854	0,856	0,858	0,860	0,862	0,864	0,866	0,868	0,870	0,872
42	0,874	0,876	0,879	0,881	0,883	0,885	0,887	0,889	0,891	0,893
43	0,895	0,897	0,899	0,901	0,904	0,906	0,908	0,910	0,912	0,914
44	0,916	0,918	0,920	0,922	0,924	0,926	0,929	0,931	0,933	0,935
45	0,937	0,939	0,941	0,943	0,945	0,947	0,949	0,951	0,954	0,956
46	0,958	0,960	0,962	0,964	0,966	0,968	0,970	0,972	0,974	0,976
47	0,979	0,981	0,983	0,985	0,987	0,989	0,991	0,993	0,995	0,997
48	0,999	1,001	1,003	1,006	1,008	1,010	1,012	1,014	1,016	1,018
49	1,020	1,022	1,024	1,026	1,028	1,031	1,033	1,035	1,037	1,039
50	1,041	1,043	1,045	1,047	1,049	1,051	1,053	1,056	1,058	1,060

Whole mg	Tenths of a mg									
	0	1	2	3	4	5	6	7	8	9
51	1,062	1,064	1,066	1,068	1,070	1,072	1,074	1,076	1,078	1,081
52	1,083	1,085	1,087	1,089	1,091	1,093	1,095	1,097	1,099	1,101
53	1,103	1,105	1,108	1,110	1,112	1,114	1,116	1,118	1,120	1,122
54	1,124	1,126	1,128	1,130	1,133	1,135	1,137	1,139	1,141	1,143
55	1,145	1,147	1,149	1,151	1,153	1,155	1,158	1,160	1,162	1,164
56	1,166	1,168	1,170	1,172	1,174	1,176	1,178	1,180	1,183	1,185
57	1,187	1,189	1,191	1,193	1,195	1,197	1,199	1,201	1,203	1,205
58	1,208	1,210	1,212	1,214	1,216	1,218	1,220	1,222	1,224	1,226
59	1,228	1,230	1,232	1,235	1,237	1,239	1,241	1,243	1,245	1,247
60	1,249	1,251	1,253	1,255	1,257	1,260	1,262	1,264	1,266	1,268
61	1,270	1,272	1,274	1,276	1,278	1,280	1,282	1,285	1,287	1,289
62	1,291	1,293	1,295	1,297	1,299	1,301	1,303	1,305	1,307	1,310
63	1,312	1,314	1,316	1,318	1,320	1,322	1,324	1,326	1,328	1,330
64	1,332	1,335	1,337	1,339	1,341	1,343	1,345	1,347	1,349	1,351
65	1,353	1,355	1,357	1,359	1,362	1,364	1,366	1,368	1,370	1,372
66	1,374	1,376	1,378	1,380	1,382	1,384	1,387	1,389	1,391	1,393
67	1,395	1,397	1,399	1,401	1,403	1,405	1,407	1,409	1,412	1,414
68	1,416	1,418	1,420	1,422	1,424	1,426	1,428	1,430	1,432	1,434
69	1,437	1,439	1,441	1,443	1,445	1,447	1,449	1,451	1,453	1,455
70	1,457	1,459	1,462	1,464	1,466	1,468	1,470	1,472	1,474	1,476
71	1,478	1,480	1,482	1,484	1,486	1,489	1,491	1,493	1,495	1,497
72	1,499	1,501	1,503	1,505	1,507	1,509	1,511	1,514	1,516	1,518
73	1,520	1,522	1,524	1,526	1,528	1,530	1,532	1,534	1,536	1,539
74	1,541	1,543	1,545	1,547	1,549	1,551	1,553	1,555	1,557	1,559
75	1,561	1,564	1,566	1,568	1,570	1,572	1,574	1,576	1,578	1,580
76	1,582	1,584	1,586	1,588	1,591	1,593	1,595	1,597	1,599	1,601
77	1,603	1,605	1,607	1,609	1,611	1,613	1,616	1,618	1,620	1,622
78	1,624	1,626	1,628	1,630	1,632	1,634	1,636	1,638	1,641	1,643
79	1,645	1,647	1,649	1,651	1,653	1,655	1,657	1,659	1,661	1,663
80	1,666	1,668	1,670	1,672	1,674	1,676	1,678	1,680	1,682	1,684
81	1,686	1,688	1,691	1,693	1,695	1,697	1,699	1,701	1,703	1,705
82	1,707	1,709	1,711	1,713	1,715	1,718	1,720	1,722	1,724	1,726
83	1,728	1,730	1,732	1,734	1,736	1,738	1,740	1,743	1,745	1,747
84	1,749	1,751	1,753	1,755	1,757	1,759	1,761	1,763	1,765	1,768
85	1,770	1,772	1,774	1,776	1,778	1,780	1,782	1,784	1,786	1,788
86	1,790	1,793	1,795	1,797	1,799	1,801	1,803	1,805	1,807	1,809
87	1,811	1,813	1,815	1,818	1,820	1,822	1,824	1,826	1,828	1,830
88	1,832	1,834	1,836	1,838	1,840	1,842	1,845	1,847	1,849	1,851
89	1,853	1,855	1,857	1,859	1,861	1,863	1,865	1,867	1,870	1,872
90	1,874	1,876	1,878	1,880	1,882	1,884	1,886	1,888	1,890	1,892
91	1,895	1,897	1,899	1,901	1,903	1,905	1,907	1,909	1,911	1,913
92	1,915	1,917	1,920	1,922	1,924	1,926	1,928	1,930	1,932	1,934
93	1,936	1,938	1,940	1,942	1,945	1,947	1,949	1,951	1,953	1,955
94	1,957	1,959	1,961	1,963	1,965	1,967	1,969	1,972	1,974	1,976
95	1,978	1,980	1,982	1,984	1,986	1,988	1,990	1,992	1,994	1,997
96	1,999	2,001	2,003	2,005	2,007	2,009	2,011	2,013	2,015	2,017
97	2,019	2,022	2,024	2,026	2,028	2,030	2,032	2,034	2,036	2,038
98	2,040	2,042	2,044	2,047	2,049	2,051	2,053	2,055	2,057	2,059
99	2,061	2,063	2,065	2,067	2,069	2,072	2,074	2,076	2,078	2,080
100	2,082	2,084	2,086	2,088	2,090	2,092	2,094	2,096	2,099	2,101

mg	1000	2000	3000	4000	5000	6000	7000	8000	9000	10 000
meq	20,82	41,64	62,46	83,28	104,10	124,91	145,73	166,55	187,37	208,19

Table 14

Conversion of Milligrams of HCO_3^- to Milligram—Equivalents
(equivalent weight of HCO_3^- = 61.019)

Whole mg	Tenths of a mg									
	0	1	2	3	4	5	6	7	8	9
0	—	0,00	0,00	0,00	0,01	0,01	0,01	0,01	0,01	0,01
1	0,016	0,018	0,020	0,021	0,023	0,025	0,026	0,028	0,030	0,031
2	0,033	0,034	0,036	0,038	0,039	0,041	0,043	0,044	0,046	0,048
3	0,049	0,051	0,052	0,054	0,056	0,057	0,059	0,061	0,062	0,064
4	0,066	0,067	0,069	0,070	0,072	0,074	0,075	0,077	0,079	0,080
5	0,082	0,084	0,085	0,087	0,089	0,090	0,092	0,093	0,095	0,097
6	0,098	0,100	0,102	0,103	0,105	0,107	0,108	0,110	0,111	0,113
7	0,115	0,116	0,118	0,120	0,121	0,123	0,125	0,126	0,128	0,129
8	0,131	0,133	0,134	0,136	0,138	0,139	0,141	0,143	0,144	0,146
9	0,148	0,149	0,151	0,152	0,154	0,156	0,157	0,159	0,161	0,162
10	0,164	0,166	0,167	0,169	0,171	0,172	0,174	0,176	0,177	0,179
11	0,180	0,182	0,184	0,185	0,187	0,189	0,190	0,192	0,193	0,195
12	0,197	0,198	0,200	0,202	0,203	0,205	0,207	0,208	0,210	0,212
13	0,213	0,215	0,216	0,218	0,220	0,221	0,223	0,225	0,226	0,228
14	0,229	0,231	0,233	0,234	0,236	0,238	0,239	0,241	0,243	0,244
15	0,246	0,248	0,249	0,251	0,253	0,254	0,256	0,257	0,259	0,261
16	0,262	0,264	0,266	0,267	0,269	0,271	0,272	0,274	0,275	0,277
17	0,279	0,280	0,282	0,284	0,285	0,287	0,289	0,290	0,292	0,294
18	0,295	0,297	0,298	0,300	0,302	0,303	0,305	0,307	0,308	0,310
19	0,311	0,313	0,315	0,316	0,318	0,320	0,321	0,323	0,325	0,326
20	0,328	0,329	0,331	0,333	0,334	0,336	0,338	0,339	0,341	0,343
21	0,344	0,346	0,348	0,349	0,351	0,352	0,354	0,356	0,357	0,359
22	0,361	0,362	0,364	0,366	0,367	0,369	0,371	0,372	0,374	0,376
23	0,377	0,379	0,380	0,382	0,384	0,385	0,387	0,389	0,390	0,392
24	0,394	0,395	0,397	0,398	0,400	0,402	0,403	0,405	0,407	0,408
25	0,410	0,411	0,413	0,415	0,416	0,418	0,420	0,421	0,423	0,425
26	0,426	0,428	0,430	0,431	0,433	0,435	0,436	0,438	0,439	0,441
27	0,443	0,444	0,446	0,448	0,449	0,451	0,452	0,454	0,456	0,457
28	0,459	0,461	0,462	0,464	0,466	0,467	0,469	0,471	0,472	0,474
29	0,475	0,477	0,479	0,480	0,482	0,484	0,485	0,487	0,488	0,490
30	0,492	0,493	0,495	0,497	0,498	0,500	0,502	0,503	0,505	0,507
31	0,508	0,510	0,511	0,513	0,515	0,516	0,518	0,520	0,521	0,523
32	0,525	0,526	0,528	0,529	0,531	0,533	0,534	0,536	0,538	0,539
33	0,541	0,543	0,544	0,546	0,548	0,549	0,551	0,552	0,554	0,556
34	0,557	0,559	0,561	0,562	0,564	0,566	0,567	0,569	0,570	0,572
35	0,574	0,575	0,577	0,579	0,580	0,582	0,584	0,585	0,587	0,589
36	0,590	0,592	0,593	0,595	0,597	0,598	0,600	0,602	0,603	0,605
37	0,606	0,608	0,610	0,611	0,613	0,615	0,616	0,618	0,620	0,621
38	0,623	0,625	0,626	0,628	0,630	0,631	0,633	0,634	0,636	0,638
39	0,639	0,641	0,643	0,644	0,646	0,648	0,649	0,651	0,652	0,654
40	0,656	0,657	0,659	0,661	0,662	0,664	0,665	0,667	0,669	0,670
41	0,672	0,674	0,675	0,677	0,679	0,680	0,682	0,684	0,685	0,687
42	0,688	0,690	0,692	0,693	0,695	0,697	0,698	0,700	0,702	0,703
43	0,704	0,706	0,708	0,710	0,711	0,713	0,715	0,716	0,718	0,720
44	0,721	0,723	0,725	0,726	0,728	0,729	0,731	0,733	0,734	0,736
45	0,738	0,739	0,741	0,743	0,744	0,746	0,747	0,749	0,751	0,752
46	0,754	0,756	0,757	0,759	0,761	0,762	0,764	0,766	0,767	0,769
47	0,770	0,772	0,774	0,775	0,777	0,779	0,780	0,782	0,784	0,785
48	0,787	0,788	0,790	0,792	0,793	0,795	0,797	0,798	0,800	0,802
49	0,803	0,805	0,807	0,808	0,810	0,811	0,813	0,815	0,816	0,818
50	0,819	0,821	0,823	0,825	0,826	0,828	0,829	0,831	0,833	0,834

Whole mg	Tenths of a mg									
	0	1	2	3	4	5	6	7	8	9
51	0,836	0,838	0,839	0,841	0,843	0,844	0,846	0,848	0,849	0,851
52	0,852	0,854	0,856	0,857	0,859	0,861	0,862	0,864	0,865	0,867
53	0,869	0,870	0,872	0,874	0,875	0,877	0,879	0,880	0,882	0,884
54	0,885	0,887	0,888	0,890	0,892	0,893	0,895	0,897	0,898	0,900
55	0,902	0,903	0,905	0,906	0,908	0,910	0,911	0,913	0,915	0,916
56	0,918	0,920	0,921	0,923	0,925	0,926	0,928	0,929	0,931	0,933
57	0,934	0,936	0,938	0,939	0,941	0,943	0,944	0,946	0,947	0,949
58	0,951	0,952	0,954	0,956	0,957	0,959	0,961	0,962	0,964	0,966
59	0,967	0,969	0,970	0,972	0,974	0,975	0,977	0,979	0,980	0,982
60	0,983	0,985	0,987	0,988	0,990	0,992	0,993	0,995	0,997	0,998
61	1,000	1,002	1,003	1,005	1,007	1,008	1,010	1,011	1,013	1,015
62	1,016	1,018	1,020	1,021	1,023	1,025	1,026	1,028	1,029	1,031
63	1,033	1,034	1,036	1,038	1,039	1,041	1,042	1,044	1,046	1,047
64	1,049	1,051	1,052	1,054	1,056	1,057	1,059	1,061	1,062	1,064
65	1,065	1,067	1,069	1,070	1,072	1,074	1,075	1,077	1,079	1,080
66	1,082	1,083	1,085	1,087	1,088	1,090	1,091	1,093	1,095	1,096
67	1,098	1,100	1,102	1,103	1,105	1,106	1,108	1,110	1,111	1,113
68	1,115	1,116	1,118	1,120	1,121	1,123	1,124	1,126	1,128	1,129
69	1,131	1,133	1,134	1,136	1,138	1,139	1,141	1,143	1,144	1,146
70	1,147	1,149	1,151	1,152	1,154	1,156	1,157	1,159	1,161	1,162
71	1,164	1,165	1,167	1,169	1,170	1,172	1,174	1,175	1,177	1,179
72	1,180	1,182	1,184	1,185	1,187	1,188	1,190	1,192	1,193	1,195
73	1,197	1,198	1,200	1,202	1,203	1,205	1,206	1,208	1,210	1,211
74	1,213	1,215	1,216	1,218	1,220	1,221	1,223	1,224	1,226	1,228
75	1,229	1,231	1,233	1,234	1,236	1,238	1,239	1,241	1,242	1,244
76	1,246	1,247	1,249	1,251	1,252	1,254	1,256	1,257	1,259	1,261
77	1,262	1,264	1,265	1,267	1,269	1,270	1,272	1,274	1,275	1,277
78	1,279	1,280	1,282	1,283	1,285	1,287	1,288	1,290	1,292	1,293
79	1,295	1,297	1,298	1,300	1,302	1,303	1,305	1,306	1,308	1,310
80	1,311	1,313	1,315	1,316	1,318	1,320	1,321	1,323	1,324	1,326
81	1,328	1,329	1,331	1,333	1,334	1,336	1,338	1,339	1,341	1,343
82	1,344	1,346	1,347	1,349	1,351	1,352	1,354	1,356	1,357	1,359
83	1,361	1,362	1,364	1,365	1,367	1,369	1,370	1,372	1,374	1,375
84	1,377	1,379	1,380	1,382	1,384	1,385	1,387	1,388	1,390	1,392
85	1,393	1,395	1,397	1,398	1,400	1,402	1,403	1,405	1,406	1,408
86	1,410	1,411	1,413	1,415	1,416	1,418	1,419	1,421	1,423	1,424
87	1,426	1,428	1,429	1,431	1,433	1,434	1,436	1,438	1,439	1,441
88	1,442	1,444	1,446	1,447	1,449	1,451	1,452	1,454	1,456	1,457
89	1,459	1,460	1,462	1,464	1,465	1,467	1,469	1,470	1,472	1,474
90	1,475	1,477	1,479	1,480	1,482	1,483	1,485	1,487	1,488	1,490
91	1,492	1,493	1,495	1,497	1,498	1,500	1,501	1,503	1,505	1,506
92	1,508	1,510	1,511	1,513	1,515	1,516	1,518	1,520	1,521	1,523
93	1,524	1,526	1,528	1,529	1,531	1,533	1,534	1,536	1,538	1,539
94	1,541	1,542	1,544	1,546	1,547	1,549	1,551	1,552	1,554	1,556
95	1,557	1,559	1,561	1,562	1,564	1,565	1,567	1,569	1,570	1,572
96	1,574	1,575	1,577	1,579	1,580	1,582	1,583	1,585	1,587	1,588
97	1,590	1,592	1,593	1,595	1,597	1,598	1,600	1,601	1,603	1,605
98	1,606	1,608	1,610	1,611	1,613	1,615	1,616	1,618	1,619	1,621
99	1,623	1,625	1,626	1,628	1,630	1,631	1,633	1,634	1,636	1,638
100	1,639	1,641	1,642	1,644	1,646	1,647	1,649	1,651	1,652	1,654

mg	1000	2000	3000	4000	5000	6000	7000	8000	9000	10 000
meq	16,39	32,78	49,17	65,55	81,94	98,33	114,72	131,11	147,50	163,89

Table 15

Conversion of Milligrams of CO_3^{2-} into Milligram—Equivalents
(equivalent weight of CO_3^{2-} = 30.0055)

Whole mg	Tenths of a mg									
	0	1	2	3	4	5	6	7	8	9
0	—	0,00	0,01	0,01	0,01	0,02	0,02	0,02	0,03	0,03
1	0,033	0,037	0,040	0,043	0,047	0,050	0,053	0,057	0,060	0,063
2	0,067	0,070	0,073	0,077	0,080	0,083	0,087	0,090	0,093	0,097
3	0,100	0,103	0,107	0,110	0,113	0,117	0,120	0,123	0,127	0,130
4	0,133	0,137	0,140	0,143	0,147	0,150	0,153	0,157	0,160	0,163
5	0,167	0,170	0,173	0,177	0,180	0,183	0,187	0,190	0,193	0,197
6	0,200	0,203	0,207	0,210	0,213	0,217	0,220	0,223	0,227	0,230
7	0,233	0,237	0,240	0,243	0,247	0,250	0,253	0,257	0,260	0,263
8	0,267	0,270	0,273	0,277	0,280	0,283	0,287	0,290	0,293	0,297
9	0,300	0,303	0,307	0,310	0,313	0,317	0,320	0,323	0,327	0,330
10	0,333	0,337	0,340	0,343	0,347	0,350	0,353	0,357	0,360	0,363
11	0,367	0,370	0,373	0,377	0,380	0,383	0,387	0,390	0,393	0,397
12	0,400	0,403	0,407	0,410	0,413	0,417	0,420	0,423	0,427	0,430
13	0,433	0,437	0,440	0,443	0,447	0,450	0,453	0,457	0,460	0,463
14	0,467	0,470	0,473	0,477	0,480	0,483	0,487	0,490	0,493	0,497
15	0,500	0,503	0,507	0,510	0,513	0,517	0,520	0,523	0,527	0,530
16	0,533	0,537	0,540	0,543	0,547	0,550	0,553	0,557	0,560	0,563
17	0,567	0,570	0,573	0,577	0,580	0,583	0,587	0,590	0,593	0,597
18	0,600	0,603	0,607	0,610	0,613	0,617	0,620	0,623	0,627	0,630
19	0,633	0,637	0,640	0,643	0,647	0,650	0,653	0,657	0,660	0,663
20	0,667	0,670	0,673	0,677	0,680	0,683	0,687	0,690	0,693	0,697
21	0,700	0,703	0,707	0,710	0,713	0,717	0,720	0,723	0,727	0,730
22	0,733	0,737	0,740	0,743	0,747	0,750	0,753	0,757	0,760	0,763
23	0,767	0,770	0,773	0,777	0,780	0,783	0,787	0,790	0,793	0,797
24	0,800	0,803	0,807	0,810	0,813	0,817	0,820	0,823	0,827	0,830
25	0,833	0,837	0,840	0,843	0,847	0,850	0,853	0,857	0,860	0,863
26	0,867	0,870	0,873	0,877	0,880	0,883	0,887	0,890	0,893	0,897
27	0,900	0,903	0,907	0,910	0,913	0,917	0,920	0,923	0,927	0,930
28	0,933	0,937	0,940	0,943	0,947	0,950	0,953	0,957	0,960	0,963
29	0,967	0,970	0,973	0,977	0,980	0,983	0,987	0,990	0,993	0,997
30	1,000	1,003	1,007	1,010	1,013	1,017	1,020	1,023	1,027	1,030
31	1,033	1,037	1,040	1,043	1,046	1,050	1,053	1,056	1,060	1,063
32	1,066	1,070	1,073	1,076	1,080	1,083	1,086	1,090	1,093	1,096
33	1,100	1,103	1,106	1,110	1,113	1,116	1,120	1,123	1,126	1,130
34	1,133	1,136	1,140	1,143	1,146	1,150	1,153	1,156	1,160	1,163
35	1,166	1,170	1,173	1,176	1,180	1,183	1,186	1,190	1,193	1,196
36	1,200	1,203	1,206	1,210	1,213	1,216	1,220	1,223	1,226	1,230
37	1,233	1,236	1,240	1,243	1,246	1,250	1,253	1,256	1,260	1,263
38	1,266	1,270	1,273	1,276	1,280	1,283	1,286	1,290	1,293	1,296
39	1,300	1,303	1,306	1,310	1,313	1,316	1,320	1,323	1,326	1,330
40	1,333	1,336	1,340	1,343	1,346	1,350	1,353	1,356	1,360	1,363
41	1,366	1,370	1,373	1,376	1,380	1,383	1,386	1,390	1,393	1,396
42	1,400	1,403	1,406	1,410	1,413	1,416	1,420	1,423	1,426	1,430
43	1,433	1,436	1,440	1,443	1,446	1,450	1,453	1,456	1,460	1,463
44	1,466	1,470	1,473	1,476	1,480	1,483	1,486	1,490	1,493	1,496
45	1,500	1,503	1,506	1,510	1,513	1,516	1,520	1,523	1,526	1,530
46	1,533	1,536	1,540	1,543	1,546	1,550	1,553	1,556	1,560	1,563
47	1,566	1,570	1,573	1,576	1,580	1,583	1,586	1,590	1,593	1,596
48	1,600	1,603	1,606	1,610	1,613	1,616	1,620	1,623	1,626	1,630
49	1,633	1,636	1,640	1,643	1,646	1,650	1,653	1,656	1,660	1,663
50	1,666	1,670	1,673	1,676	1,680	1,683	1,686	1,690	1,693	1,696

Table 16

Conversion of Milligrams of NO$_3^-$ into Milligram—Equivalents
(equivalent weight of NO$_3^-$ = 62.008.)

Whole mg	Tenths of a mg									
	0	1	2	3	4	5	6	7	8	9
0	—	0,00	0,00	0,00	0,01	0,01	0,01	0,01	0,01	0,01
1	0,016	0,018	0,019	0,021	0,023	0,024	0,025	0,027	0,029	0,031
2	0,032	0,034	0,035	0,037	0,039	0,040	0,042	0,044	0,045	0,047
3	0,048	0,050	0,052	0,053	0,055	0,056	0,058	0,060	0,061	0,063
4	0,065	0,066	0,068	0,070	0,071	0,072	0,074	0,076	0,077	0,079
5	0,081	0,082	0,084	0,085	0,087	0,089	0,090	0,092	0,094	0,095
6	0,097	0,098	0,100	0,102	0,103	0,105	0,106	0,108	0,110	0,111
7	0,113	0,115	0,116	0,118	0,119	0,121	0,123	0,124	0,126	0,127
8	0,129	0,131	0,132	0,134	0,135	0,137	0,139	0,140	0,142	0,144
9	0,145	0,147	0,148	0,150	0,152	0,153	0,155	0,156	0,158	0,160
10	0,161	0,163	0,164	0,166	0,168	0,169	0,171	0,173	0,174	0,176
11	0,177	0,179	0,181	0,182	0,184	0,185	0,187	0,189	0,190	0,192
12	0,194	0,195	0,197	0,198	0,200	0,202	0,203	0,205	0,206	0,208
13	0,210	0,211	0,213	0,214	0,216	0,218	0,219	0,221	0,223	0,224
14	0,226	0,227	0,229	0,231	0,232	0,234	0,235	0,237	0,239	0,240
15	0,242	0,244	0,245	0,247	0,248	0,250	0,252	0,253	0,255	0,256
16	0,258	0,260	0,261	0,263	0,264	0,266	0,268	0,269	0,271	0,273
17	0,274	0,276	0,277	0,279	0,281	0,282	0,284	0,285	0,287	0,289
18	0,290	0,292	0,294	0,295	0,297	0,298	0,300	0,302	0,303	0,305
19	0,306	0,308	0,310	0,311	0,313	0,314	0,316	0,318	0,319	0,321
20	0,323	0,324	0,326	0,327	0,329	0,331	0,332	0,334	0,335	0,337
21	0,339	0,340	0,342	0,344	0,345	0,347	0,348	0,350	0,352	0,353
22	0,355	0,356	0,358	0,360	0,361	0,363	0,364	0,366	0,368	0,369
23	0,371	0,373	0,374	0,376	0,377	0,379	0,381	0,382	0,384	0,3 5
24	0,387	0,389	0,390	0,392	0,393	0,395	0,397	0,398	0,400	0,402
25	0,403	0,405	0,406	0,408	0,410	0,411	0,413	0,414	0,416	0,418
26	0,419	0,421	0,423	0,424	0,426	0,427	0,429	0,431	0,432	0,434
27	0,435	0,437	0,439	0,440	0,442	0,443	0,445	0,447	0,448	0,450
28	0,452	0,453	0,455	0,456	0,458	0,460	0,461	0,463	0,464	0,466
29	0,468	0,469	0,471	0,473	0,474	0,476	0,477	0,479	0,481	0,482
30	0,484	0,485	0,487	0,489	0,490	0,492	0,493	0,495	0,497	0,498
31	0,500	0,502	0,503	0,505	0,506	0,508	0,510	0,511	0,513	0,514
32	0,516	0,518	0,519	0,521	0,523	0,524	0,526	0,527	0,529	0,531
33	0,532	0,534	0,535	0,537	0,539	0,540	0,542	0,543	0,545	0,547
34	0,548	0,550	0,552	0,553	0,555	0,556	0,558	0,560	0,561	0,563
35	0,564	0,566	0,568	0,569	0,571	0,572	0,574	0,576	0,577	0,579
36	0,581	0,582	0,584	0,585	0,587	0,589	0,590	0,592	0,593	0,595
37	0,597	0,598	0,600	0,602	0,603	0,605	0,606	0,608	0,610	0,611
38	0,613	0,614	0,616	0,618	0,619	0,621	0,622	0,624	0,626	0,627
39	0,629	0,631	0,632	0,634	0,635	0,637	0,639	0,640	0,642	0,643
40	0,645	0,647	0,648	0,650	0,652	0,653	0,655	0,656	0,658	0,660
41	0,661	0,663	0,664	0,666	0,668	0,669	0,671	0,672	0,674	0,676
42	0,677	0,679	0,681	0,682	0,684	0,685	0,687	0,689	0,690	0,692
43	0,693	0,695	0,697	0,698	0,700	0,702	0,703	0,705	0,706	0,708
44	0,710	0,711	0,713	0,714	0,716	0,718	0,719	0,721	0,722	0,724
45	0,726	0,727	0,729	0,731	0,732	0,734	0,735	0,737	0,739	0,740
46	0,742	0,743	0,745	0,747	0,748	0,750	0,752	0,753	0,755	0,756
47	0,758	0,760	0,761	0,763	0,764	0,766	0,768	0,769	0,771	0,772
48	0,774	0,776	0,777	0,779	0,781	0,782	0,784	0,785	0,787	0,789
49	0,790	0,792	0,793	0,795	0,797	0,798	0,800	0,802	0,803	0,804
50	0,806	0,808	0,810	0,811	0,813	0,814	0,816	0,818	0,819	0,821

Table 17

Conversion of Milligrams of NO_2^- into Milligram—Equivalents
(equivalent weight of $NO_2^- = 46.008$)

Whole mg	Tenths of a mg									
	0	1	2	3	4	5	6	7	8	9
0	—	0,00	0,00	0,01	0,01	0,01	0,01	0,02	0,02	0,02
1	0,022	0,024	0,026	0,028	0,030	0,033	0,035	0,037	0,039	0,041
2	0,043	0,046	0,048	0,050	0,052	0,054	0,057	0,059	0,061	0,063
3	0,065	0,067	0,070	0,072	0,074	0,076	0,078	0,080	0,083	0,085
4	0,087	0,089	0,091	0,093	0,096	0,098	0,100	0,102	0,104	0,107
5	0,109	0,111	0,113	0,115	0,117	0,120	0,122	0,124	0,126	0,128
6	0,130	0,133	0,135	0,137	0,139	0,141	0,143	0,146	0,148	0,150
7	0,152	0,154	0,157	0,159	0,161	0,163	0,165	0,167	0,170	0,172
8	0,174	0,176	0,178	0,180	0,183	0,185	0,187	0,189	0,191	0,193
9	0,196	0,198	0,200	0,202	0,204	0,206	0,209	0,211	0,213	0,215
10	0,217	0,220	0,222	0,224	0,226	0,228	0,230	0,233	0,235	0,237

Table 18

Conversion of Milligrams of Br^- into Milligram—Equivalents
(equivalent weight of $Br^- = 79.916$)

Whole mg	Tenths of a mg									
	0	1	2	3	4	5	6	7	8	9
0	—	0,00	0,00	0,00	0,00	0,01	0,01	0,01	0,01	0,01
1	0,013	0,014	0,015	0,016	0,018	0,019	0,020	0,021	0,023	0,024
2	0,025	0,026	0,028	0,029	0,030	0,031	0,033	0,034	0,035	0,036
3	0,038	0,039	0,040	0,041	0,043	0,044	0,045	0,046	0,048	0,049
4	0,050	0,051	0,053	0,054	0,055	0,056	0,058	0,059	0,060	0,061
5	0,063	0,064	0,065	0,066	0,068	0,069	0,070	0,071	0,073	0,074
6	0,075	0,076	0,078	0,079	0,080	0,081	0,083	0,084	0,085	0,086
7	0,088	0,089	0,090	0,091	0,093	0,094	0,095	0,096	0,098	0,099
8	0,100	0,101	0,103	0,104	0,105	0,106	0,108	0,109	0,110	0,111
9	0,113	0,114	0,115	0,116	0,118	0,119	0,120	0,121	0,123	0,124
10	0,125	0,126	0,128	0,129	0,130	0,131	0,133	0,134	0,135	0,136

mg	20	30	40	50	60	70	80	90	100
meq	0,250	0,375	0,500	0,625	0,751	0,876	1,001	1,126	1,251

Table 19

Conversion of Milligrams of I⁻ into Milligram—Equivalents
(equivalent weight of I⁻ = 126.91)

Whole mg	Tenths of a mg									
	0	1	2	3	4	5	6	7	8	9
0	—	0,00	0,00	0,00	0,00	0,00	0,00	0,01	0,01	0,01
1	0,008	0,009	0,009	0,010	0,011	0,012	0,013	0,013	0,014	0,015
2	0,016	0,016	0,017	0,018	0,019	0,020	0,020	0,021	0,022	0,023
3	0,024	0,024	0,025	0,026	0,027	0,027	0,028	0,029	0,030	0,031
4	0,032	0,032	0,033	0,034	0,035	0,035	0,036	0,037	0,038	0,039
5	0,039	0,040	0,041	0,042	0,043	0,043	0,044	0,045	0,046	0,046
6	0,047	0,048	0,049	0,050	0,050	0,051	0,052	0,053	0,054	0,054
7	0,055	0,056	0,057	0,058	0,058	0,059	0,060	0,061	0,061	0,062
8	0,063	0,064	0,065	0,065	0,066	0,067	0,068	0,069	0,069	0,070
9	0,071	0,072	0,072	0,073	0,074	0,075	0,076	0,076	0,077	0,078
10	0,079	0,080	0,080	0,081	0,082	0,083	0,084	0,084	0,085	0,086

Table 20

Conversion of Milligrams of F⁻ into Milligram—Equivalents
(equivalent weight of F⁻ = 19.00)

Whole mg	Tenths of a mg									
	0	1	2	3	4	5	6	7	8	9
0	—	0,01	0,01	0,02	0,02	0,03	0,03	0,04	0,04	0,05
1	0,053	0,058	0,063	0,068	0,074	0,079	0,084	0,089	0,095	0,100
2	0,105	0,111	0,116	0,121	0,126	0,132	0,137	0,142	0,147	0,153
3	0,158	0,163	0,168	0,174	0,179	0,184	0,189	0,195	0,200	0,205
4	0,211	0,216	0,221	0,226	0,232	0,237	0,242	0,247	0,253	0,258
5	0,263	0,268	0,274	0,279	0,284	0,289	0,295	0,300	0,305	0,311
6	0,316	0,321	0,326	0,332	0,337	0,342	0,347	0,353	0,358	0,363
7	0,368	0,374	0,379	0,384	0,389	0,395	0,400	0,405	0,411	0,416
8	0,421	0,426	0,432	0,437	0,442	0,447	0,453	0,458	0,463	0,468
9	0,474	0,479	0,484	0,489	0,495	0,500	0,505	0,511	0,516	0,521
10	0,526	0,532	0,537	0,542	0,547	0,553	0,558	0,563	0,568	0,574

IV. Tables for Converting Milligram—Equivalents into Milligrams

Table 21

Conversion of Milligram—Equivalents of Na$^+$ into Milligrams
(equivalent weight of Na$^+$ = 22.991)

Whole and tenths of meq	Hundredths of a meq									
	0	1	2	3	4	5	6	7	8	9
0,0	—	0,2	0,5	0,7	0,9	1,1	1,4	1,6	1,8	2,1
0,1	2,30	2,53	2,76	2,99	3,22	3,45	3,68	3,91	4,14	4,37
0,2	4,60	4,83	5,06	5,29	5,52	5,75	5,98	6,21	6,44	6,67
0,3	6,90	7,13	7,36	7,59	7,82	8,05	8,28	8,51	8,74	8,97
0,4	9,20	9,43	9,66	9,89	10,12	10,35	10,58	10,81	11,04	11,27
0,5	11,50	11,73	11,96	12,19	12,42	12,65	12,87	13,10	13,33	13,56
0,6	13,79	14,02	14,25	14,48	14,71	14,94	15,17	15,40	15,63	15,86
0,7	16,09	16,32	16,55	16,78	17,01	17,24	17,47	17,70	17,93	18,16
0,8	18,39	18,62	18,85	19,08	19,31	19,54	19,77	20,00	20,23	20,46
0,9	20,69	20,92	21,15	21,38	21,61	21,84	22,07	22,30	22,53	22,76
1,0	22,99	23,22	23,45	23,68	23,91	24,14	24,37	24,60	24,83	25,06
1,1	25,29	25,52	25,75	25,98	26,21	26,44	26,67	26,90	27,13	27,36
1,2	27,59	27,82	28,05	28,28	28,51	28,74	28,97	29,20	29,43	29,66
1,3	29,89	30,12	30,35	30,58	30,81	31,04	31,27	31,50	31,73	31,96
1,4	32,19	32,42	32,65	32,88	33,11	33,34	33,57	33,80	34,03	34,26
1,5	34,49	34,72	34,95	35,18	35,41	35,64	35,87	36,10	36,33	36,56
1,6	36,79	37,02	37,25	37,48	37,71	37,94	38,16	38,39	38,62	38,85
1,7	39,08	39,31	39,54	39,77	40,00	40,23	40,46	40,69	40,92	41,15
1,8	41,38	41,61	41,84	42,07	42,30	42,53	42,76	42,99	43,22	43,45
1,9	43,68	43,91	44,14	44,37	44,60	44,83	45,06	45,29	45,52	45,75
2,0	45,98	46,21	46,44	46,67	46,90	47,13	47,36	47,59	47,82	48,05
2,1	48,28	48,51	48,74	48,97	49,20	49,43	49,66	49,89	50,12	50,35
2,2	50,58	50,81	51,04	51,27	51,50	51,73	51,96	52,19	52,42	52,65
2,3	52,88	53,11	53,34	53,57	53,80	54,03	54,26	54,49	54,72	54,95
2,4	55,18	55,41	55,64	55,87	56,10	56,33	56,56	56,79	57,02	57,25
2,5	57,48	57,71	57,94	58,17	58,40	58,63	58,86	59,09	59,32	59,55
2,6	59,78	60,01	60,24	60,47	60,70	60,93	61,16	61,39	61,62	61,85
2,7	62,07	62,30	62,53	62,76	62,99	63,22	63,45	63,69	63,91	64,14
2,8	64,37	64,60	64,83	65,06	65,29	65,52	65,75	65,98	66,21	66,44
2,9	66,67	66,90	67,13	67,36	67,59	67,82	68,05	68,28	68,51	68,74
3,0	68,97	69,20	69,43	69,66	69,89	70,12	70,35	70,58	70,81	71,04
3,1	71,27	71,50	71,73	71,96	72,19	72,42	72,65	72,88	73,11	73,34
3,2	73,57	73,80	74,03	74,26	74,49	74,72	74,95	75,18	75,41	75,64
3,3	75,87	76,10	76,33	76,56	76,79	77,02	77,25	77,48	77,71	77,94
3,4	78,17	78,40	78,63	78,86	79,09	79,32	79,55	79,78	80,01	80,24
3,5	80,47	80,70	80,93	81,16	81,39	81,62	81,85	82,08	82,31	82,54
3,6	82,77	83,00	83,23	83,46	83,69	83,92	84,15	84,38	84,61	84,84
3,7	85,07	85,30	85,53	85,76	85,99	86,22	86,45	86,68	86,91	87,14
3,8	87,37	87,60	87,83	88,06	88,29	88,52	88,75	88,98	89,21	89,43
3,9	89,66	89,89	90,12	90,35	90,58	90,81	91,01	91,27	91,50	91,73
4,0	91,96	92,19	92,42	92,65	92,88	93,11	93,34	93,57	93,80	94,03
4,1	94,26	94,49	94,72	94,95	95,18	95,41	95,64	95,87	96,10	96,33
4,2	96,56	96,79	97,02	97,25	97,48	97,71	97,94	98,17	98,40	98,63
4,3	98,86	99,09	99,32	99,55	99,78	100,01	100,24	100,47	100,70	100,93
4,4	101,16	101,39	101,62	101,85	102,08	102,31	102,54	102,77	103,00	103,23
4,5	103,46	103,69	103,92	104,15	104,38	104,61	104,84	105,07	105,30	105,53
4,6	105,76	105,99	106,22	106,45	106,68	106,91	107,14	107,37	107,60	107,83
4,7	108,06	108,29	108,52	108,75	108,98	109,21	109,44	109,67	109,90	110,13
4,8	110,36	110,59	110,82	111,05	111,28	111,51	111,74	111,97	112,20	112,43
4,9	112,66	112,89	113,12	113,35	113,58	113,81	114,04	114,27	114,50	114,73
5,0	114,96	115,18	115,41	115,64	115,87	116,10	116,33	116,56	116,79	117,02

Whole and tenths of meq	Hundredths of a meq									
	0	1	2	3	4	5	6	7	8	9
5,1	117,25	117,48	117,71	117,94	118,17	118,40	118,63	118,86	119,09	119,32
5,2	119,55	119,78	120,01	120,24	120,47	120,70	120,93	121,16	121,39	121,62
5,3	121,85	122,08	122,31	122,54	122,77	123,00	123,23	123,46	123,69	123,92
5,4	124,15	124,38	124,61	124,84	125,07	125,30	125,53	125,76	125,99	126,22
5,5	126,45	126,68	126,91	127,14	127,37	127,60	127,83	128,06	128,29	128,52
5,6	128,75	128,98	129,21	129,44	129,67	129,90	130,13	130,36	130,59	130,82
5,7	131,05	131,28	131,51	131,74	131,97	132,20	132,43	132,66	132,89	133,12
5,8	133,35	133,58	133,81	134,04	134,27	134,50	134,73	134,96	135,19	135,42
5,9	135,65	135,88	136,11	136,34	136,57	136,80	137,03	137,26	137,49	137,72
6,0	137,95	138,18	138,41	138,64	138,87	139,10	139,33	139,56	139,79	140,02
6,1	140,25	140,48	140,70	140,93	141,16	141,39	141,62	141,85	142,08	142,31
6,2	142,54	142,77	143,00	143,23	143,46	143,69	143,92	144,15	144,38	144,61
6,3	144,84	145,07	145,30	145,53	145,76	145,99	146,22	146,45	146,68	146,91
6,4	147,14	147,37	147,60	147,83	148,06	148,29	148,52	148,75	148,98	149,21
6,5	149,44	149,67	149,90	150,13	150,36	150,59	150,82	151,05	151,28	151,51
6,6	151,74	151,97	152,20	152,43	152,66	152,89	153,12	153,35	153,58	153,81
6,7	154,04	154,27	154,50	154,73	154,96	155,19	155,42	155,65	155,88	156,11
6,8	156,34	156,57	156,80	157,03	157,26	157,49	157,72	157,95	158,18	158,41
6,9	158,64	158,87	159,10	159,33	159,56	159,79	160,02	160,25	160,48	160,71
7,0	160,94	161,17	161,40	161,63	161,86	162,09	162,32	162,55	162,78	163,01
7,1	163,24	163,47	163,70	163,93	164,16	164,39	164,62	164,85	165,08	165,31
7,2	165,54	165,77	166,00	166,22	166,45	166,68	166,91	167,14	167,37	167,60
7,3	167,83	168,06	168,29	168,52	168,75	168,98	169,21	169,44	169,67	169,90
7,4	170,13	170,36	170,59	170,82	171,05	171,28	171,51	171,74	171,97	172,20
7,5	172,43	172,66	172,89	173,12	173,35	173,58	173,81	174,04	174,27	174,50
7,6	174,73	174,96	175,19	175,42	175,65	175,88	176,11	176,34	176,57	176,80
7,7	177,03	177,26	177,49	177,72	177,95	178,18	178,41	178,64	178,87	179,10
7,8	179,33	179,56	179,79	180,02	180,25	180,48	180,71	180,94	181,17	181,40
7,9	181,63	181,86	182,09	182,32	182,55	182,78	183,01	183,24	183,47	183,70
8,0	183,93	184,16	184,39	184,62	184,85	185,08	185,31	185,54	185,77	186,00
8,1	186,23	186,46	186,69	186,92	187,15	187,38	187,61	187,84	188,07	188,30
8,2	188,53	188,76	188,99	189,22	189,45	189,68	189,91	190,14	190,37	190,60
8,3	190,83	191,06	191,29	191,52	191,74	191,97	192,20	192,43	192,66	192,89
8,4	193,12	193,35	193,58	193,81	194,04	194,27	194,50	194,73	194,96	195,19
8,5	195,42	195,65	195,88	196,11	196,34	196,57	196,80	197,03	197,26	197,49
8,6	197,72	197,95	198,18	198,41	198,64	198,87	199,10	199,33	199,56	199,79
8,7	200,02	200,25	200,48	200,71	200,94	201,17	201,40	201,63	201,86	202,09
8,8	202,32	202,55	202,78	203,01	203,24	203,47	203,70	203,93	204,16	204,39
8,9	204,62	204,85	205,08	205,31	205,54	205,77	206,00	206,23	206,46	206,69
9,0	206,92	207,15	207,38	207,61	207,84	208,07	208,30	208,53	208,76	208,99
9,1	209,22	209,45	209,68	209,91	210,14	210,37	210,60	210,83	211,06	211,29
9,2	211,52	211,75	211,98	212,21	212,44	212,67	212,90	213,13	213,36	213,59
9,3	213,82	214,05	214,28	214,51	214,74	214,97	215,20	215,43	215,66	215,89
9,4	216,12	216,35	216,58	216,81	217,04	217,26	217,49	217,72	217,95	218,18
9,5	218,41	218,64	218,87	219,10	219,33	219,56	219,79	220,02	220,25	220,48
9,6	220,71	220,94	221,17	221,40	221,63	221,86	222,09	222,32	222,55	222,78
9,7	223,01	223,24	223,47	223,70	223,93	224,16	224,39	224,62	224,85	225,08
9,8	225,31	225,54	225,77	226,00	226,23	226,46	226,69	226,92	227,15	227,38
9,9	227,61	227,84	228,07	228,30	228,53	228,76	228,99	229,22	229,45	229,68
10,0	229,91	230,14	230,37	230,60	230,83	231,06	231,29	231,52	231,75	231,98

meq	100	200	300	400	500	600	700	800	900	1000
mg	2299,1	4598,2	6897,3	9196,4	11495,5	13794,6	16093,7	18392,8	20691,9	22991

Table 22

Conversion of Milligram-Equivalents of Ca^{2+} into Milligrams
(equivalent weight of Ca^{2+} = 20.04)

Whole and tenths of meq	Hundredths of a meq									
	0	1	2	3	4	5	6	7	8	9
0,0	—	0,2	0,4	0,6	0,8	1,0	1,2	1,4	1,6	1,8
0,1	2,00	2,20	2,40	2,61	2,81	3,01	3,21	3,41	3,61	3,81
0,2	4,01	4,21	4,41	4,61	4,81	5,01	5,21	5,41	5,61	5,81
0,3	6,01	6 21	6,41	6,61	6,81	7,01	7,21	7,41	7,62	7,82
0,4	8,02	8,22	8,42	8,62	8,82	9,02	9,22	9,42	9,62	9,82
0,5	10,02	10,22	10,42	10,62	10,82	11,02	11,22	11,42	11,62	11,82
0,6	12,02	12,22	12,42	12,63	12,83	13,03	13,23	13,43	13,63	13,83
0,7	14,03	14,23	14,43	14,63	14,83	15,03	15,23	15,43	15,63	15,83
0,8	16,03	16,23	16,43	16,63	16,83	17,03	17,23	17,43	17,64	17,84
0,9	18,04	18,24	18,44	18,64	18,84	19,04	19,24	19,44	19,64	19,84
1,0	20,04	20,24	20,44	20,64	20,84	21,04	21,24	21,44	21,64	21,84
1,1	22,04	22,24	22,44	22,65	22,85	23,05	23,25	23,45	23,65	23,85
1,2	24,05	24,25	24,45	24,65	24,85	25,05	25,25	25,45	25,65	25,85
1,3	26,05	26,25	26,45	26,65	26,85	27,05	27,25	27,45	27,66	27,86
1,4	28,06	28,26	28,46	28,66	28,86	29,06	29,26	29,46	29,66	29,86
1,5	30,06	30,26	30,46	30,66	30,86	31,06	31,26	31,46	31,66	31,86
1,6	32,06	32,26	32,46	32,67	32,87	33,07	33,27	33,47	33,67	33,87
1,7	34,07	34,27	34,47	34,67	34,87	35,07	35,27	35,47	35,67	35,87
1,8	36,07	36,27	36,47	36,67	36,87	37,07	37,27	37,47	37,68	37,88
1,9	38,08	38,28	38,48	38,68	38,88	39,08	39,28	39,48	39,68	39,88
2,0	40,08	40,28	40,48	40,68	40,88	41,08	41,28	41,48	41,68	41,88
2,1	42,08	42,28	42,48	42,69	42,89	43,09	43,29	43,49	43,69	43,89
2,2	44,09	44,29	44,49	44,69	44,89	45,09	45,29	45,49	45,69	45,89
2,3	46,09	46,29	46,49	46,69	46,89	47,09	47,29	47,49	47,70	47,90
2,4	48,10	48,30	48,50	48,70	48,90	49,10	49,30	49,50	49,70	49,90
2,5	50,10	50,30	50,50	50,70	50,90	51,10	51,30	51,50	51,70	51,90
2,6	52,10	52,30	52,50	52,71	52,91	53,11	53,31	53,51	53,71	53,91
2,7	54,11	54,31	54,51	54,71	54,91	55,11	55,31	55,51	55,71	55,91
2,8	56,11	56,31	56,51	56,71	56,91	57,11	57,31	57,51	57,72	57,92
2,9	58,12	58,32	58,52	58,72	58,92	59,12	59,32	59,52	59,72	59,92
3,0	60,12	60,32	60,52	60,72	60,92	61,12	61,32	61,52	61,72	61,92
3,1	62,12	62,32	62,52	62,73	62,93	63,13	63,33	63,53	63,73	63,93
3,2	64,13	64,33	64,53	64,73	64,93	65,13	65,33	65,53	65,73	65,93
3,3	66,13	66,33	66,53	66,73	66,93	67,13	67,33	67,53	67,74	67,94
3,4	68,14	68,34	68,54	68,74	68,94	69,14	69,34	69,54	69,74	69,94
3,5	70,14	70,34	70,54	70,74	70,94	71,14	71,34	71,54	71,74	71,94
3,6	72,14	72,34	72,54	72,75	72,95	73,15	73,35	73,55	73,75	73,95
3,7	74,15	74,35	74,55	74,75	74,95	75,15	75,35	75,55	75,75	75,95
3,8	76,15	76,35	76,55	76,75	76,95	77,15	77,35	77,55	77,76	77,96
3,9	78,16	78,36	78,56	78,76	78,96	79,16	79,36	79,56	79,76	79,96
4,0	80,16	80,36	80,56	80,76	80,96	81,16	81,36	81,56	81,76	81,96
4,1	82,16	82,36	82,56	82,77	82,97	83,17	83,37	83,57	83,77	83,97
4,2	84,17	84,37	84,57	84,77	84,97	85,17	85,37	85,57	85,77	85,97
4,3	86,17	86,37	86,57	86,77	86,97	87,17	87,37	87,57	87,78	87,98
4,4	88,18	88,38	88,58	88,78	88,98	89,18	89,38	89,58	89,78	89,98
4,5	90,18	90,38	90,58	90,78	90,98	91,18	91,38	91,58	91,78	91,98
4,6	92,18	92,38	92,58	92,79	92,99	93,19	93,39	93,59	93,79	93,99
4,7	94,19	94,39	94,59	94,79	94,99	95,19	95,39	95,59	95,79	95,99
4,8	96,19	96,39	96,59	96,79	96,99	97,19	97,39	97,59	97,80	98,00
4,9	98,20	98,40	98,60	98,80	99,00	99,20	99,40	99,60	99,80	100,00
5,0	100,20	100,40	100,60	100,80	101,00	101,20	101,40	101,60	101,80	102,00

Whole and tenths of meq	Hundredths of a meq									
	0	1	2	3	4	5	6	7	8	9
5,1	102,20	102,40	102,60	102,81	103,01	103,21	103,41	103,61	103,81	104,01
5,2	104,21	104,41	104,61	104,81	105,01	105,21	105,41	105,61	105,81	106,01
5,3	106,21	106,41	106,61	106,81	107,01	107,21	107,41	107,61	107,82	108,02
5,4	108,22	108,42	108,62	108,82	109,02	109,22	109,42	109,62	109,82	110,02
5,5	110,22	110,42	110,62	110,82	111,02	111,22	111,42	111,62	111,82	112,02
5,6	112,22	112,42	112,62	112,83	113,03	113,23	113,43	113,63	113,83	114,03
5,7	114,23	114,43	114,63	114,83	115,03	115,23	115,43	115,63	115,83	116,03
5,8	116,23	116,43	116,63	116,83	117,03	117,23	117,43	117,63	117,84	118,04
5,9	118,24	118,44	118,64	118,84	119,04	119,24	119,44	119,64	119,84	120,04
6,0	120,24	120,44	120,64	120,84	121,04	121,24	121,44	121,64	121,84	122,04
6,1	122,24	122,44	122,64	122,85	123,05	123,25	123,45	123,65	123,85	124,05
6,2	124,25	124,45	124,65	124,85	125,05	125,25	125,45	125,65	125,85	126,05
6,3	126,25	126,45	126,65	126,85	127,05	127,25	127,45	127,65	127,86	128,06
6,4	128,26	128,46	128,66	128,86	129,06	129,26	129,46	129,66	129,86	130,06
6,5	130,26	130,46	130,66	130,86	131,06	131,26	131,46	131,66	131,86	132,06
6,6	132,26	132,46	132,66	132,87	133,07	133,27	133,47	133,67	133,87	134,07
6,7	134,27	134,47	134,67	134,87	135,07	135,27	135,47	135,67	135,87	136,07
6,8	136,27	136,47	136,67	136,87	137,07	137,27	137,47	137,67	137,88	138,08
6,9	138,28	138,48	138,68	138,88	139,08	139,28	139,48	139,68	139,88	140,08
7,0	140,28	140,48	140,68	140,88	141,08	141,28	141,48	141,68	141,88	142,08
7,1	142,28	142,48	142,68	142,89	143,09	143,29	143,49	143,69	143,89	144,09
7,2	144,29	144,49	144,69	144,89	145,09	145,29	145,49	145,69	145,89	146,09
7,3	146,29	146,49	146,69	146,89	147,09	147,29	147,49	147,69	147,90	148,10
7,4	148,30	148,50	148,70	148,90	149,10	149,30	149,50	149,70	149,90	150,10
7,5	150,30	150,50	150,70	150,90	151,10	151,30	151,50	151,70	151,90	152,10
7,6	152,30	152,50	152,70	152,91	153,11	153,31	153,51	153,71	153,91	154,11
7,7	154,31	154,51	154,71	154,91	155,11	155,31	155,51	155,71	155,91	156,11
7,8	156,31	156,51	156,71	156,91	157,11	157,31	157,51	157,71	157,92	158,12
7,9	158,32	158,52	158,72	158,92	159,12	159,32	159,52	159,72	159,92	160,12
8,0	160,32	160,52	160,72	160,92	161,12	161,32	161,52	161,72	161,92	162,12
8,1	162,32	162,52	162,72	162,93	163,13	163,33	163,53	163,73	163,93	164,13
8,2	164,33	164,53	164,73	164,93	165,13	165,33	165,53	165,73	165,93	166,13
8,3	166,33	166,53	166,73	166,93	167,13	167,33	167,53	167,73	167,94	168,14
8,4	168,34	168,54	168,74	168,94	169,14	169,34	169,54	169,74	169,94	170,14
8,5	170,34	170,54	170,74	170,94	171,14	171,34	171,54	171,74	171,94	172,14
8,6	172,34	172,54	172,74	172,95	173,15	173,35	173,55	173,75	173,95	174,15
8,7	174,35	174,55	174,75	174,95	175,15	175,35	175,55	175,75	175,95	176,15
8,8	176,35	176,55	176,75	176,95	177,15	177,35	177,55	177,75	177,96	178,16
8,9	178,36	178,56	178,76	178,96	179,16	179,36	179,56	179,76	179,96	180,16
9,0	180,36	180,56	180,76	180,96	181,16	181,36	181,56	181,76	181,96	182,16
9,1	182,36	182,56	182,76	182,97	183,17	183,37	183,57	183,77	183,97	184,17
9,2	184,37	184,57	184,77	184,97	185,17	185,37	185,57	185,77	185,97	186,17
9,3	186,37	186,57	186,77	186,97	187,17	187,37	187,57	187,77	187,98	188,18
9,4	188,38	188,58	188,78	188,98	189,18	189,38	189,58	189,78	189,98	190,18
9,5	190,38	190,58	190,78	190,98	191,18	191,38	191,58	191,78	191,98	192,18
9,6	192,38	192,58	192,78	192,99	193,19	193,39	193,59	193,79	193,99	194,19
9,7	194,39	194,59	194,79	194,99	195,19	195,39	195,59	195,79	195,99	196,19
9,8	196,39	196,59	196,79	196,99	197,19	197,39	197,59	197,79	198,00	198,20
9,9	198,40	198,60	198,80	199,00	199,20	199,40	199,60	199,80	200,00	200,20
10,0	200,40	200,60	200,80	201,00	201,20	201,40	201,60	201,80	202,00	202,20

meq	100	200	300	400	500	600	700	800	900	1000
mg	2004,0	4008,0	6012,0	8016,0	10020,0	12024,0	14028,0	16032,0	18036,0	20040

Table 23

Conversion of Milligram—Equivalents of Mg^{2+} into Milligrams
(equivalent weight of Mg^{2+} = 12.16)

Whole and tenths of meq	Hundredths of a meq									
	0	1	2	3	4	5	6	7	8	9
0,0	—	0,1	0,2	0,4	0,5	0,6	0,7	0,9	1,0	1,1
0,1	1,22	1,34	1,46	1,58	1,70	1,82	1,95	2,07	2,19	2,31
0,2	2,43	2,55	2,68	2,80	2,92	3,04	3,16	3,28	3,40	3,53
0,3	3 65	3,77	3,89	4,01	4,13	4,26	4,38	4,50	4.62	4,74
0,4	4,86	4,99	5,11	5,23	5,35	5,47	5,59	5,72	5,84	5,96
0,5	6,08	6,20	6,32	6,44	6,57	6,69	6,81	6,93	7,05	7,17
0,6	7,30	7,42	7,54	7,66	7,78	7,90	8,03	8,15	8,27	8,39
0,7	8,51	8,63	8,76	8,88	9,00	9,12	9,24	9,36	9,48	9,61
0,8	9,73	9,85	9,97	10,09	10,21	10,34	10,46	10,58	10,70	10,82
0,9	10,94	11,07	11,19	11,31	11,43	11,55	11,67	11,80	11,92	12,04
1,0	12.16	12,28	12,40	12,52	12,65	12,77	12,89	13,01	13,13	13,25
1,1	13,38	13,50	13,62	13,74	13,86	13,98	14,11	14,23	14,35	14,47
1,2	14,59	14,71	14,84	14,96	15,08	15,20	15,32	15,44	15,56	15,69
1,3	15,81	15,93	16,05	16,17	16,29	16,42	16,54	16,66	16,78	16,90
1,4	17,02	17,15	17,27	17,39	17,51	17,63	17,75	17,88	18,00	18,12
1,5	18,24	18,36	18,48	18,60	18,73	18,85	18,97	19,09	19,21	19,33
1,6	19,46	19,58	19,70	19,82	19,94	20,06	20,19	20,31	20,43	20,55
1,7	20,67	20,79	20,92	21,04	21,16	21,28	21,40	21,52	2i,64	21,77
1,8	21,89	22,01	22,13	22,25	22,37	22,50	22,62	22,74	22,86	22,98
1,9	23,10	23,23	23,35	23,47	23,59	23,71	23,83	23,96	24,08	24,20
2,0	24,32	24,44	24,56	24,68	24,81	24,93	25,05	25,17	25,29	25,41
2,1	25,54	25 66	25,78	25,90	26,02	26,14	26,27	26,39	26,51	26,63
2,2	26,75	26,87	27,00	27,12	27,24	27,36	27,48	27,60	27,72	27,85
2,3	27,97	28,09	28,21	28,33	28,45	28,58	28,70	28,82	28,94	29,06
2,4	29,18	29,31	29,43	29,55	29,67	29,79	29,91	30,04	30,16	30,28
2,5	30,40	30,52	30,64	30,76	30,89	31,01	31,13	31,25	31,37	31,49
2,6	31,62	31,74	31,86	31,98	32,10	32,22	32,35	32,47	32,59	32,71
2,7	32,83	32,95	33,08	33,20	33,32	33,44	33,56	33,68	33,80	33,93
2,8	34,05	34,17	34,29	34,41	34,53	34,66	34,78	34,90	35,02	35,14
2,9	35,26	35,39	35,51	35,63	35,75	35.87	35,99	36,12	36,24	36,36
3,0	36,48	36,60	36,72	36.84	36,97	37,09	37,21	37,33	37,45	37,57
3,1	37,70	37,82	37,94	38,06	38,18	38,30	38,43	38,55	38,67	38,79
3,2	38,91	39,03	39,16	39,28	39,40	39,52	39,64	39,76	39,88	40,01
3,3	40,13	40,25	40,37	40,49	40,61	40,74	40,86	40,98	41,10	41,22
3,4	41,34	41,47	41,59	41,71	41,83	41,95	42,07	42,20	42,32	42,44
3,5	42,56	42,68	42,80	42,92	43,05	43,17	43,29	43,41	43,53	43,65
3,6	43,78	43,90	44,02	44,14	44,26	44,38	44,51	44,63	44,75	44,87
3,7	44,99	45,11	45,24	45,36	45,48	45,60	45,72	45,84	45,96	46,09
3,8	46,21	46,33	46,45	46,57	46,69	46,82	46,94	47,06	47,18	47,30
3,9	47,42	47,55	47,67	47,79	47,91	48,03	48,15	48,28	48,40	48,52
4,0	48.64	48 76	48,88	49,00	49,13	49,25	49,37	49,49	49,61	49,73
4,1	49,86	49,98	50,10	50,22	50,34	50,46	50,59	50,71	50,83	50,95
4,2	51,07	51,19	51,32	51,44	51,56	51,68	51,80	51,92	52,04	52,17
4,3	52,29	52,41	52,53	52,65	52,77	52,90	53,02	53,14	53,26	53,38
4,4	53,50	53,63	53,75	53,87	53,99	54,11	54,23	54,36	54,48	54,60
4,5	54,72	54,84	54,96	55,08	55,21	55,33	55,45	55,57	55,69	55,81
4,6	55,94	56,06	56,18	56,30	56,42	56,54	56,67	56,79	56,91	57,03
4,7	57,15	57,27	57,40	57,52	57,64	57,76	57,88	58,00	58,12	58,25
4,8	58,37	58,49	58,61	58,73	58,85	58,98	59,10	59,22	59,34	59,46
4,9	59,58	59,71	59,83	59,95	60,07	60,19	60,31	60,44	60,56	60,68
5,0	60,80	60,92	61,04	61,16	61,29	61,41	61,53	61,65	61,77	61,89

Whole and tenths of meq	Hundredths of a meq									
	0	1	2	3	4	5	6	7	8	9
5,1	62,02	62,14	62,26	62,38	62,50	62,62	62,75	62,87	62,99	63,11
5,2	63,23	63,35	63,48	63,60	63,72	63,84	63,96	64,08	64,20	64,33
5,3	64,45	64,57	64,69	64,81	64,93	65,06	65,18	65,30	65,42	65,54
5,4	65,66	65,79	65,91	66,03	66,15	66,27	66,39	66,52	66,64	66,76
5,5	66,88	67,00	67,12	67,24	67,37	67,49	67,61	67,73	67,85	67,97
5,6	68,10	68,22	68,34	68,46	68,58	68,70	68,83	68,95	69,07	69,19
5,7	69,31	69,43	69,56	69,68	69,80	69,92	70,04	70,16	70,28	70,41
5,8	70,53	70,65	70,77	70,89	71,01	71,14	71,26	71,38	71,50	71,62
5,9	71,74	71,87	71,99	72,11	72,23	72,35	72,47	72,60	72,72	72,84
6,0	72,96	73,08	73,20	73,32	73,45	73,57	73,69	73,81	73,93	74,05
6,1	74,18	74,30	74,42	74,54	74,66	74,78	74,91	75,03	75,15	75,27
6,2	75,39	75,51	75,64	75,76	75,88	76,00	76,12	76,24	76,36	76,49
6,3	76,61	76,73	76,85	76,97	77,09	77,22	77,34	77,46	77,58	77,70
6,4	77,82	77,95	78,07	78,19	78,31	78,43	78,55	78,68	78,80	78,92
6,5	79,04	79,16	79,28	79,40	79,53	79,65	79,77	79,89	80,01	80,13
6,6	80,26	80,38	80,50	80,62	80,74	80,86	80,99	81,11	81,23	81,35
6,7	81,47	81,59	81,72	81,84	81,96	82,08	82,20	82,32	82,44	82,57
6,8	82,69	82,81	82,93	83,05	83,17	83,30	83,42	83,54	83,66	83,78
6,9	83,90	84,03	84,15	84,27	84,39	84,51	84,63	84,76	84,88	85,00
7,0	85,12	85,24	85,36	85,48	85,61	85,73	85,85	85,97	86,09	86,21
7,1	86,34	86,46	86,58	86,70	86,82	86,94	87,07	87,19	87,31	87,43
7,2	87,55	87,67	87,80	87,92	88,04	88,16	88,28	88,40	88,52	88,65
7,3	88,77	88,89	89,01	89,13	89,25	89,38	89,50	89,62	89,74	89,86
7,4	89,98	90,11	90,23	90,35	90,47	90,59	90,71	90,84	90,96	91,08
7,5	91,20	91,32	91,44	91,56	91,69	91,81	91,93	92,05	92,17	92,29
7,6	92,42	92,54	92,66	92,78	92,90	93,02	93,15	93,27	93,39	93,51
7,7	93,63	93,75	93,88	94,00	94,12	94,24	94,36	94,48	94,60	94,73
7,8	94,85	94,97	95,09	95,21	95,33	95,46	95,58	95,70	95,82	95,94
7,9	96,06	96,19	96,31	96,43	96,55	96,67	96,79	96,92	97,04	97,16
8,0	97,28	97,40	97,52	97,64	97,77	97,89	98,01	98,13	98,25	98,37
8,1	98,50	98,62	98,74	98,86	98,98	99,10	99,23	99,35	99,47	99,59
8,2	99,71	99,83	99,96	100,08	100,20	100,32	100,44	100,56	100,68	100,81
8,3	100,93	101,05	101,17	101,29	101,41	101,54	101,66	101,78	101,90	102,02
8,4	102,14	102,27	102,39	102,51	102,63	102,75	102,87	103,00	103,12	103,24
8,5	103,36	103,48	103,60	103,72	103,85	103,97	104,09	104,21	104,33	104,45
8,6	104,58	104,70	104,82	104,94	105,06	105,18	105,31	105,43	105,55	105,67
8,7	105,79	105,91	106,04	106,16	106,28	106,40	106,52	106,64	106,76	106,89
8,8	107,01	107,13	107,25	107,37	107,49	107,62	107,74	107,86	107,98	108,10
8,9	108,22	108,35	108,47	108,59	108,71	108,83	108,95	109,08	109,20	109,32
9,0	109,44	109,56	109,68	109,80	109,93	110,05	110,17	110,29	110,41	110,53
9,1	110,66	110,78	110,90	111,02	111,14	111,26	111,39	111,51	111,63	111,75
9,2	111,87	111,99	112,12	112,24	112,36	112,48	112,60	112,72	112,84	112,97
9,3	113,09	113,21	113,33	113,45	113,57	113,70	113,82	113,94	114,06	114,18
9,4	114,30	114,43	114,55	114,67	114,79	114,91	115,03	115,16	115,28	115,40
9,5	115,52	115,64	115,76	115,88	116,01	116,13	116,25	116,37	116,49	116,61
9,6	116,74	116,86	116,98	117,10	117,22	117,34	117,47	117,59	117,71	117,83
9,7	117,95	118,07	118,20	118,32	118,44	118,56	118,68	118,80	118,92	119,05
9,8	119,17	119,29	119,41	119,53	119,65	119,78	119,90	120,02	120,14	120,26
9,9	120,38	120,51	120,63	120,75	120,87	120,99	121,11	121,24	121,36	121,48
10,0	121,60	121,72	121,84	121,96	122,09	122,21	122,33	122,45	122,57	122,69

meq	100	200	300	400	500	600	700	800	900	1000
mg	1216,0	2432,0	3648,0	4864,0	6080,0	7296,0	8512,0	9728,0	10944,0	12160

Table 24

Conversion of Milligram—Equivalents of K^+ into Milligrams
(equivalent weight of K^+ = 39.100)

Whole and tenths of meq	Hundredths of a meq									
	0	1	2	3	4	5	6	7	8	9
0,0	—	0,4	0,8	1,2	1,6	2,0	2,3	2,7	3,1	3,5
0,1	3,91	4,30	4,69	5,08	5,47	5,86	6,26	6,65	7,04	7,43
0,2	7,82	8,21	8,60	8,99	9,38	9,77	10,16	10,56	10,95	11,34
0,3	11,73	12,12	12,51	12,90	13,29	13,68	14,07	14,47	14,86	15,25
0,4	15,64	16,03	16,42	16,81	17,20	17,59	17,98	18,38	18,77	19,16
0,5	19,55	19,94	20,33	20,72	21,11	21,50	21,89	22,28	22,68	23,07
0,6	23,46	23,85	24,24	24,63	25,02	25,41	25,80	26,19	26,59	26,98
0,7	27,37	27,76	28,15	28,54	28,93	29,32	29,71	30,10	30,49	30,89
0,8	31,28	31,67	32,06	32,45	32,84	33,23	33,62	34,01	34,40	34,80
0,9	35,19	35,58	35,97	36,36	36,75	37,14	37,53	37,92	38,31	38,71
1,0	39,10	39,49	39,88	40,27	40,66	41,05	41,44	41,83	42,22	42,61
1,1	43,01	43,40	43,79	44,18	44,57	44,96	45,35	45,74	46,13	46,52
1,2	46,92	47,31	47,70	48,09	48,48	48,87	49,27	49,66	50,05	50,44
1,3	50,83	51,23	51,62	52,01	52,40	52,79	53,18	53,57	53,96	54,35
1,4	54,74	55,14	55,53	55,92	56,31	56,70	57,09	57,48	57,87	58,26
1,5	58,65	59,04	59,44	59,83	60,22	60,61	61,00	61,39	61,78	62,17
1,6	62,56	62,95	63,35	63,74	64,13	64,52	64,91	65,30	65,69	66,08
1,7	66,47	66,86	67,26	67,65	68,04	68,43	68,82	69,21	69,60	69,99
1,8	70,38	70,77	71,16	71,56	71,95	72,34	72,73	73,12	73,51	73,90
1,9	74,29	74,68	75,07	75,47	75,86	76,25	76,64	77,03	77,42	77,81
2,0	78,20	78,59	78,98	79,37	79,77	80,16	80,55	80,94	81,33	81,72
2,1	82,11	82,50	82,89	83,28	83,68	84,07	84,46	84,85	85,24	85,63
2,2	86,02	86,41	86,80	87,19	87,59	87,98	88,37	88,76	89,15	89,54
2,3	89,93	90,32	90,71	91,10	91,49	91,89	92,28	92,67	93,06	93,45
2,4	93,84	94,23	94,62	95,01	95,40	95,80	96,19	96,58	96,97	97,36
2,5	97,75	98,14	98,53	98,92	99,31	99,70	100,10	100,49	100,88	101,27
2,6	101,66	102,05	102,44	102,83	103,22	103,61	104,01	104,40	104,79	105,18
2,7	105,57	105,96	106,35	106,74	107,13	107,52	107,91	108,29	108,70	109,09
2,8	109,48	109,87	110,26	110,65	111,04	111,43	111,82	112,22	112,61	113,00
2,9	113,39	113,78	114,17	114,56	114,95	115,34	115,73	116,13	116,52	116,91
3,0	117,30	117,69	118,08	118,47	118,86	119,25	119,64	120,03	120,43	120,82
3,1	121,21	121,60	121,99	122,38	122,77	123,16	123,55	123,94	124,34	124,73
3,2	125,12	125,51	125,90	126,29	126,68	127,07	127,46	127,85	128,24	128,64
3,3	129,03	129,42	129,81	130,30	130,59	130,98	131,37	131,76	132,15	132,55
3,4	132,94	133,33	133,72	134,11	134,50	134,89	135,28	135,67	136,06	136,46
3,5	136,85	137,24	137,63	138,02	138,41	138,80	139,19	139,58	139,97	140,36
3,6	140,76	141,15	141,54	141,93	142,32	142,71	143,10	143,49	143,88	144,27
3,7	144,67	145,06	145,45	145,84	146,23	146,62	147,02	147,41	147,80	148,19
3,8	148,58	148,98	149,37	149,76	150,15	150,54	150,93	151,32	151,71	152,10
3,9	152,49	152,89	153,28	153,67	154,06	154,45	154,84	155,23	155,62	156,01
4,0	156,40	156,79	157,19	157,58	157,97	158,36	158,75	159,14	159,53	159,92
4,1	160,31	160,70	161,10	161,49	161,88	162,27	162,66	163,05	163,44	163,83
4,2	164,22	164,61	165,01	165,40	165,79	166,18	166,57	166,96	167,35	167,74
4,3	168,13	168,52	168,91	169,31	169,70	170,09	170,48	170,87	171,26	171,65
4,4	172,04	172,43	172,82	173,22	173,61	174,00	174,39	174,78	175,17	175,56
4,5	175,95	176,34	176,73	177,12	177,52	177,91	178,30	178,69	179,08	179,47
4,6	179,86	180,25	180,64	181,03	181,43	181,82	182,21	182,60	182,99	183,38
4,7	183,77	184,16	184,55	184,94	185,34	185,72	186,12	186,51	186,90	187,29
4,8	187,68	188,07	188,46	188,85	189,24	189,64	190,03	190,42	190,81	191,20
4,9	191,59	191,98	192,37	192,76	193,15	193,55	193,94	194,32	194,72	195,11
5,0	195,50	195,89	196,28	196,67	197,06	197,45	197,85	198,24	198,63	199,02

Table 25

Conversion of Milligram—Equivalents of Fe^{2+} into Milligrams
(equivalent weight of Fe^{2+} = 27.925)

Whole and tenths of meq	Hundredths of a meq									
	0	1	2	3	4	5	6	7	8	9
0,0	—	0,3	0,6	0,8	1,1	1,4	1,7	2,0	2,2	2,5
0,1	2,79	3,07	3,35	3,63	3,91	4,19	4,47	4,75	5,03	5,31
0,2	5,59	5,86	6,14	6,42	6,70	6,98	7,26	7,54	7,82	8,10
0,3	8,38	8,66	8,94	9,22	9,49	9,77	10,05	10,33	10,61	10,89
0,4	11,17	11,45	11,73	12,01	12,29	12,57	12,85	13,12	13,40	13,68
0,5	13,96	14,24	14,52	14,80	15,08	15,36	15,64	15,92	16,20	16,48
0,6	16,76	17,03	17,31	17,59	17,87	18,15	18,43	18,71	18,99	19,27
0,7	19,55	19,83	20,11	20,39	20,66	20,94	21,22	21,50	21,78	22,06
0,8	22,34	22,62	22,90	23,18	23,46	23,74	24,02	24,29	24,57	24,85
0,9	25,13	25,41	25,69	25,97	26,25	26,53	26,81	27,09	27,37	27,65
1,0	27,93	28,20	28,48	28,76	29,04	29,32	29,60	29,88	30,16	30,44
1,1	30,72	31,00	31,28	31,56	31,83	32,11	32,39	32,67	32,95	33,23
1,2	33,51	33,79	34,07	34,35	34,63	34,91	35,19	35,46	35,74	36,02
1,3	36,30	36,58	36,86	37,14	37,42	37,70	37,98	38,26	38,54	38,82
1,4	39,10	39,37	39,65	39,93	40,21	40,49	40,77	41,05	41,33	41,61
1,5	41,89	42,17	42,45	42,73	43,00	43,28	43,56	43,84	44,12	44,40
1,6	44,68	44,96	45,24	45,52	45,80	46,08	46,36	46,63	46,91	47,19
1,7	47,47	47,75	48,03	48,31	48,59	48,87	49,15	49,43	49,71	49,99
1,8	5C,27	50,54	50,82	51,10	51,38	51,66	51,94	52,22	52,50	52,78
1,9	53,06	53,34	53,62	53,90	54,17	54,45	54,73	55,01	55,29	55,57
2,0	55,85	56,13	56,41	56,69	56,97	57,25	57,53	57,80	58,08	58,36
2,1	58,64	58,92	59,20	59,48	59,76	60,04	60,32	60,60	60,88	61,16
2,2	61,44	61,71	61,99	62,27	62,55	62,83	63,11	63,40	63,67	63,95
2,3	64,23	64,51	64,79	65,07	65,34	65,62	65,90	66,18	66,46	66,74
2,4	67,02	67,30	67,58	67,86	68,14	68,42	68,70	68,97	69,25	69,53
2,5	69,81	70,09	70,37	70,65	70,93	71,21	71,49	71,77	72,05	72,33
2,6	72,61	72,88	73,16	73,44	73,72	74,00	74,28	74,56	74,84	75,12
2,7	75,40	75,68	75,96	76,24	76,51	76,79	77,07	77,35	77,63	77,91
2,8	78,19	78,47	78,75	79,03	79,31	79,59	79,87	80,14	80,42	80,70
2,9	80,98	81,26	81,54	81,82	82,10	82,38	82,66	82,94	83,22	83,50
3,0	83,78	84,05	84,33	84,61	84,89	85,17	85,45	85,73	86,01	86,29
3,1	86,57	86,85	87,13	87,41	87,68	87,96	88,24	88,52	88,80	89,08
3,2	89,36	89,64	89,92	90,20	90,48	90,76	91,04	91,31	91,59	91,87
3,3	92,15	92,43	92,71	92,99	93,27	93,55	93,83	94,11	94,39	94,67
3,4	94,95	95,22	95,50	95,78	96,06	96,34	96,62	96,90	97,18	97,46
3,5	97,74	98,02	98,30	98,58	98,85	99,13	99,41	99,69	99,97	100,25
3,6	100,53	100,81	101,09	101,37	101,65	101,93	102,21	102,48	102,76	103,04
3,7	103,32	103,60	103,88	104,16	104,44	104,72	105,00	105,28	105,56	105,84
3,8	106,12	106,39	106,67	106,95	107,23	107,51	107,79	108,07	108,35	108,63
3,9	108,91	109,19	109,47	109,75	110,02	110,30	110,58	110,86	111,14	111,42
4,0	111,70	111,98	112,26	112,54	112,82	113,10	113,38	113,65	113,93	114,21
4,1	114,49	114,77	115,05	115,33	115,61	115,89	116,17	116,45	116,73	117,01
4,2	117,29	117,56	117,84	118,12	118,40	118,68	118,96	119,24	119,52	119,80
4,3	120,08	120,36	120,64	120,92	121,19	121,47	121,75	122,03	122,31	122,59
4,4	122,87	123,15	123,43	123,71	123,99	124,27	124,55	124,82	125,10	125,38
4,5	125,66	125,94	126,22	126,50	126,78	127,06	127,34	127,62	127,90	128,18
4,6	128,46	128,73	129,01	129,29	129,57	129,85	130,13	130,41	130,69	130,97
4,7	131,25	131,53	131,81	132,09	132,36	132,64	132,92	133,20	133,48	133,76
4,8	134,04	134,32	134,60	134,88	135,16	135,44	135,72	135,99	136,27	136,55
4,9	136,83	137,11	137,39	137,67	137,95	138,23	138,51	138,79	139,07	139,35
5,0	139,63	139,90	140,18	140,46	140,74	141,02	141,30	141,58	141,86	142,14

Table 26

Conversion of Milligram—Equivalents of Fe^{3+} into Milligrams
(equivalent weight of Fe^{3+} = 18.617)

Whole and tenths of meq	Hundredths of a meq									
	0	1	2	3	4	5	6	7	8	9
0,0	—	0,2	0,4	0,6	0,7	0,9	1,1	1,3	1,5	1,7
0,1	1,86	2,05	2,23	2,42	2,61	2,79	2,98	3,16	3,35	3,54
0,2	3,72	3,91	4,10	4,28	4,47	4,65	4,84	5,03	5,21	5,40
0,3	5,59	5,77	5,96	6,14	6,33	6,52	6,70	6,89	7,07	7,26
0,4	7,45	7,63	7,82	8,01	8,19	8,38	8,56	8,75	8,94	9,12
0,5	9,31	9,49	9,68	9,87	10,05	10,24	10,43	10,61	10,80	10,98
0,6	11,17	11,36	11,54	11,73	11,91	12,10	12,29	12,47	12,66	12,85
0,7	13,03	13,22	13,40	13,59	13,77	13,96	14,15	14,34	14,52	14,71
0,8	14,89	15,08	15,27	15,45	15,64	15,82	16,01	16,20	16,38	16,57
0,9	16,76	16,94	17,13	17,31	17,50	17,69	17,87	18,06	18,24	18,43
1,0	18,62	18,80	18,99	19,18	19,36	19,55	19,73	19,92	20,11	20,29

Table 27

Conversion of Milligram—Equivalents of Al^{3+} into Milligrams
(equivalent weight of Al^{3+} = 8.993)

Whole and tenths of meq	Hundreths of a meq									
	0	1	2	3	4	5	6	7	8	9
0,0	—	0,1	0,2	0,3	0,4	0,4	0,5	0,6	0,7	0,8
0,1	0,90	0,99	1,08	1,17	1,26	1,35	1,44	1,53	1,62	1,71
0,2	1,80	1,89	1,98	2,07	2,16	2,25	2,34	2,43	2,52	2,61
0,3	2,70	2,79	2,88	2,97	3,06	3,15	3,24	3,33	3,42	3,51
0,4	3,60	3,69	3,78	3,87	3,96	4,05	4,14	4,23	4,32	4,41
0,5	4,50	4,58	4,67	4,76	4,85	4,94	5,03	5,12	5,21	5,30
0,6	5,39	5,48	5,57	5,66	5,75	5,84	5,93	6,02	6,11	6,20
0,7	6,29	6,38	6,47	6,56	6,65	6,74	6,83	6,92	7,01	7,10
0,8	7,19	7,28	7,37	7,46	7,55	7,64	7,73	7,82	7,91	8,00
0,9	8,09	8,18	8,27	8,36	8,45	8,54	8,63	8,72	8,81	8,90
1,0	8,99	9,08	9,17	9,26	9,35	9,44	9,53	9,62	9,71	9,80

Table 28

Conversion of Milligram—Equivalents of Mn^{2+} into Milligrams
(equivalent weight of Mn^{2+} = 27.47)

Whole and tenths of meq	Hundredths of a meq									
	0	1	2	3	4	5	6	7	8	9
0,0	—	0,3	0,5	0,8	1,1	1,4	1,6	1,9	2,2	2,5
0,1	2,75	3,02	3,30	3,57	3,85	4,12	4,40	4,67	4,94	5,22
0,2	5,49	5,77	6,04	6,32	6,59	6,87	7,14	7,42	7,69	7,97
0,3	8,24	8,52	8,79	9,07	9,34	9,61	9,89	10,16	10 44	10,71
0,4	10,99	11,26	11,54	11,81	12,09	12,36	12,64	12,91	13,19	13,46
0,5	13,73	14,01	14,28	14,56	14,83	15,11	15,38	15,66	15 93	16,21
0,6	16,48	16,76	17,03	17,31	17,58	17,86	18,13	18 40	18,68	18,95
0,7	19,23	19,50	19,78	20,05	20,33	20,60	20,88	21,15	21,43	21,70
0,8	21,98	22,26	22,53	22,80	23,07	23,35	23,62	23,90	24,17	24,45
0,9	24,72	25,00	25,27	25,55	25,82	26,10	26,37	26,65	26,92	27,20
1,0	27,47	27,74	28,02	28,29	28,57	28,84	29,12	29,39	29,67	29,94

Table 29

Conversion of Milligram—Equivalents of NH_4^+ into Milligrams
(equivalent weight of NH_4^+ = 18.040)

Whole and tenths of meq	Hundredths of a meq									
	0	1	2	3	4	5	6	7	8	9
0,0	—	0,2	0,4	0,5	0,7	0,9	1,1	1,3	1,4	1,6
0,1	1,80	1,98	2,16	2,35	2,53	2,71	2,89	3,07	3,25	3,43
0,2	3,61	3,79	3,97	4,15	4,33	4,51	4,69	4,87	5,05	5,23
0,3	5,41	5,59	5,77	5,95	6,13	6,31	6,49	6,67	6,86	7,04
0,4	7,22	7,40	7,58	7,76	7,94	8,12	8,30	8,48	8,66	8,84
0,5	9,02	9,20	9,38	9,56	9,74	9,92	10,10	10,28	10,46	10,64
0,6	10,82	11,00	11,18	11,37	11,55	11,73	11,91	12,09	12,27	12,45
0,7	12,63	12,81	12 99	13,17	13,35	13,53	13,71	13,89	14,07	14,25
0,8	14,43	14,61	14,79	14,97	15,15	15,33	15,51	15,69	15,88	16,06
0,9	16,24	16,42	16,60	16,78	16,96	17,14	17,32	17,50	17,68	17,86
1,0	18,04	18,22	18,40	18,58	18,76	18,94	19,12	19,30	19,48	19,66

Table 30

Conversion of Milligram—Equivalents of Cl⁻ into Milligrams
(equivalent weight of Cl⁻ = 35.457)

Whole and tenths of meq	Hundredths of a meq									
	0	1	2	3	4	5	6	7	8	9
0,0	—	0,4	0,7	1,1	1,4	1,8	2,1	2,5	2,8	3,2
0,1	3.55	3,90	4,25	4,61	4,96	5,32	5,67	6,03	6,38	6,74
0,2	7,09	7,45	7,80	8,16	8,51	8,86	9,22	9,57	9,93	10,28
0,3	10,64	10,99	11,35	11,70	12,06	12,41	12,76	13,12	13,47	13,83
0,4	14,18	14,54	14,89	15,25	15,60	15,96	16,31	16,66	17,02	17,37
0,5	17,73	18,08	18,44	18,79	19,15	19,50	19,86	20,21	20,57	20,92
0,6	21,27	21,63	21,98	22,34	22,69	23,05	23,40	23,76	24,11	24,47
0,7	24,82	25,17	25,53	25,88	26,24	26,59	26,95	27,30	27,66	28,01
0,8	28,37	28,72	29,07	29,43	29,78	30,14	30,49	30,85	31,20	31,56
0,9	31,91	32,27	32,62	32,98	33,33	33,68	34,04	34,39	34,75	35,10
1,0	35,46	35,81	36,17	36,52	36,88	37,23	37,58	37,94	38,29	38,65
1,1	39,00	39,36	39,71	40,07	40,42	40,78	41,13	41,48	41,84	42,19
1,2	42,55	42,90	43,26	43,61	43,97	44,32	44,68	45,03	45,38	45,74
1,3	46,09	46,45	46,80	47,16	47,51	47,87	48,22	48,58	48,93	49,29
1,4	49,64	49,99	50,35	50,70	51,06	51,41	51,77	52,12	52,48	52,83
1,5	53,19	53,54	53,89	54,25	54,60	54,96	55,31	55,67	56,02	56,38
1,6	56,73	57,09	57,44	57,79	58,15	58,50	58,86	59,21	59,57	59,92
1,7	60,28	60,63	60,99	61,34	61,70	62,05	62,40	62,76	63,11	63,47
1,8	63,82	64,18	64,53	64,89	65,24	65,60	65,95	66,30	66,66	67,01
1,9	67,37	67,72	68,08	68,43	68,79	69,14	69,50	69,85	70,20	70,56
2,0	70,91	71,27	71,62	71,98	72,33	72,69	73,04	73,40	73,75	74,11
2,1	74,46	74,81	75,17	75,52	75,88	76,23	76,59	76,94	77,30	77,65
2,2	78,01	78,36	78,71	79,07	79,42	79,78	80,13	80,49	80,84	81,20
2,3	81,55	81,91	82,26	82,61	82,97	83,32	83,68	84,03	84,39	84,74
2,4	85,10	85,45	85,81	86,16	86,52	86,87	87,22	87,58	87,93	88,29
2,5	88,64	89,00	89,35	89,71	90,06	90,42	90,77	91,12	91,48	91,83
2,6	92,19	92,54	92,90	93,25	93,61	93,96	94,32	94,67	95,02	95,38
2,7	95,73	96,09	96,44	96,80	97,15	97,51	97,86	98,22	98,57	98,93
2,8	99,28	99,63	99,99	100,34	100,70	101,05	101,41	101,76	10',12	102,47
2,9	102,83	103,18	103,53	103,89	104,24	104,60	104,95	105,31	105,66	106,02
3,0	106,37	106,73	107,08	107,43	107,79	108,14	108,50	108,85	109,21	109,56
3,1	109,92	110,27	110,63	110,98	111,33	111,69	112,04	112,40	112,75	113,11
3,2	113,46	113,82	114,17	114,53	114,88	115,24	115,59	115,94	116,30	116,65
3,3	117,01	117,36	117,72	118,07	118,43	118,78	119,14	119,49	119,84	120,20
3,4	120,55	120,91	121,26	121,62	121,97	122,33	122,68	123,04	123,39	123,74
3,5	124,10	124,45	124,81	125,16	125,52	125,87	126,23	126,58	126,94	127,29
3,6	127,65	128,00	128,35	128,71	129,06	129,42	129,77	130,13	130,48	130,84
3,7	131,19	131,55	131,90	132,25	132,61	132,96	133,32	133,67	134,03	134,38
3,8	134,74	135,09	135,45	135,80	136,15	136,51	136,86	137,22	137,57	137,93
3,9	138,28	138,64	138,99	139,35	139,70	140,06	140,41	140,76	141,12	141,47
4,0	141,83	142,18	142,54	142,89	143,25	143,60	143,96	144,31	144,66	145,02
4,1	145,37	145,73	146,08	146,44	146,79	147,15	147,50	147,86	148,21	148,56
4,2	148,92	149,27	149,63	149,98	150,34	150,69	151,05	151,40	151,76	152,11
4,3	152,47	152,82	153,17	153,53	153,88	154,24	154,59	154,95	155,30	155,66
4,4	156,01	156,37	156,72	157,07	157,43	157,78	158,14	158,49	158,85	159,20
4,5	159,56	159,91	160,27	160,62	160,97	161,33	161,68	162,04	162,39	162,75
4,6	163,10	163,46	163,81	164,17	164,52	164,88	165,23	165,58	165,94	166,29
4,7	166,65	167,00	167,36	167,71	168,07	168,42	168,78	169,13	169,48	169,84
4,8	170,19	170,55	170,90	171,26	171,61	171,97	172,32	172,68	173,03	173,38
4,9	173,74	174,09	174,45	174,80	175,16	175,51	175,87	176,22	176,58	176,93
5,0	177,29	177,64	177,99	178,35	178,70	179,06	179,41	179,77	180,12	180,48

Whole and tenths of meq	Hundredths of a meq									
	0	1	2	3	4	5	6	7	8	9
5,1	180,83	181,19	181,54	181,89	182,25	182,60	182,96	183,31	183,67	184,02
5,2	184,38	184,73	185,09	185,44	185,79	186,15	186,50	186,96	187,21	187,57
5,3	187,92	188,28	188,63	188,99	189,34	189,69	190,05	190,40	190,76	191,11
5,4	191,47	191,82	192,18	192,53	192,89	193,24	193,60	193,95	194,30	194,66
5,5	195,01	195,37	195,72	196,08	196,43	196,79	197,14	197,50	197,85	198,20
5,6	198,56	198,91	199,27	199,62	199,98	200,33	200,69	201,04	201,40	201,75
5,7	202,10	202,46	202,81	203,17	203,52	203,88	204,23	204,59	204,94	205,30
5,8	205,65	206,01	206,36	206,71	207,07	207,42	207,78	208,13	208,49	208,84
5,9	209,20	209,55	209,91	210,26	210,61	210,97	211,32	211,68	212,03	212,39
6,0	212,74	213,10	213,45	213,81	214,16	214,51	214,87	215,22	215,58	215,93
6,1	216,29	216,64	217,00	217,35	217,71	218,06	218,42	218,77	219,12	219,48
6,2	219,83	220,19	220,54	220,90	221,25	221,61	221,96	222,32	222,67	223,02
6,3	223,38	223,73	224,09	224,44	224,80	225,15	225,51	225,86	226,22	226,57
6,4	226,92	227,28	227,63	227,99	228,34	228,70	229,05	229,41	229,76	230,12
6,5	230,47	230,83	231,18	231,53	231,89	232,24	232,60	232,95	233,31	233,66
6,6	234,02	234,37	234,73	235,08	235,43	235,79	236,14	236,50	236,85	237,21
6,7	237,56	237,92	238,27	238,63	238,98	239,33	239,69	240,04	240,40	240,75
6,8	241,11	241,46	241,82	242,17	242,53	242,88	243,24	243,59	243,94	244,30
6,9	244,65	245,01	245,36	245,72	246,07	246,43	246,78	247,14	247,49	247,84
7,0	248,20	248,55	248,91	249,26	249,62	249,97	250,33	250,68	251,04	251,39
7,1	251,74	252,10	252,45	252,81	253,16	253,52	253,87	254,23	254,58	254,94
7,2	255,29	255,64	256,00	256,35	256,71	257,06	257,42	257,77	258,13	258,48
7,3	258,84	259,19	259,55	259,90	260,25	260,61	260,96	261,32	261,67	262,03
7,4	262,38	262,74	263,09	263,45	263,80	264,15	264,51	264,86	265,22	265,57
7,5	265,93	266,28	266,64	266,99	267,35	267,70	268,05	268,41	268,76	269,12
7,6	269,47	269,83	270,18	270,54	270,89	271,25	271,60	271,96	272,31	272,66
7,7	273,02	273,37	273,73	274,08	274,44	274,79	275,15	275,50	275,86	276,21
7,8	276,56	276,92	277,27	277,63	277,98	278,34	278,69	279,05	279,40	279,76
7,9	280,11	280,46	280,82	281,17	281,53	281,88	282,24	282,59	282,95	283,30
8,0	283,66	284,01	284,37	284,72	285,07	285,43	285,78	286,14	286,49	286,85
8,1	287,20	287,56	287,91	288,27	288,62	288,97	289,33	289,68	290,04	290,39
8,2	290,75	291,10	291,46	291,81	292,17	292,52	292,87	293,23	293,58	293,94
8,3	294,29	294,65	295,00	295,36	295,71	296,07	296,42	296,78	297,13	297,48
8,4	297,84	298,19	298,55	298,90	299,26	299,61	299,97	300,32	300,68	301,03
8,5	301,38	301,74	302,09	302,45	302,80	303,16	303,51	303,87	304,22	304,58
8,6	304,93	305,28	305,64	305,99	306,35	306,70	307,06	307,41	307,77	308,12
8,7	308,48	308,83	309,19	309,54	309,89	310,25	310,60	310,96	311,31	311,67
8,8	312,02	312,38	312,73	313,09	313,44	313,79	314,15	314,50	314,86	315,21
8,9	315,57	315,92	316,28	316,63	316,99	317,34	317,69	318,05	318,40	318,76
9,0	319,11	319,47	319,82	320,18	320,53	320,89	321,24	321,59	321,95	322,30
9,1	322,66	323,01	323,37	323,72	324,08	324,43	324,79	325,14	325,50	325,85
9,2	326,20	326,56	326,91	327,27	327,62	327,98	328,33	328,69	329,04	329,40
9,3	329,75	330,10	330,46	330,81	331,17	331,52	331,88	332,23	332,59	332,94
9,4	333,30	333,65	334,00	334,36	334,71	335,07	335,42	335,78	336,13	336,49
9,5	336,84	337,20	337,55	337,91	338,26	338,61	338,97	339,32	339,68	340,03
9,6	340,39	340,74	341,10	341,45	341,81	342,16	342,51	342,87	343,22	343,58
9,7	343,93	344,29	344,64	345,00	345,35	345,71	346,06	346,42	346,77	347,12
9,8	347,48	347,83	348,19	348,54	348,90	349,25	349,61	349,96	350,32	350,67
9,9	351,02	351,38	351,73	352,09	352,44	352,80	353,15	353,51	353,86	354,22
10,0	354,57	354,92	355,28	355,63	355,99	356,34	356,70	357,05	357,41	357,76

meq	100	200	300	400	500	600	700	800	900	1000
mg	3545,7	7091,4	10637,1	14182,8	17728,5	21274,2	24819,9	28365,6	31911,3	35457

Table 31

Conversion of Milligram—Equivalents of SO_4^{2-} into Milligrams
(equivalent weight of SO_4^{2-} = 48.033)

Whole and tenths of meq	Hundredths of a meq									
	0	1	2	3	4	5	6	7	8	9
0,0	—	0,5	1,0	1,4	1,9	2,4	2,9	3,4	3,8	4,3
0,1	4,80	5,28	5,76	6,24	6,72	7,20	7,69	8,17	8,65	9,13
0,2	9 61	10,09	10,57	11,05	11,53	12,01	12,49	12,97	13,45	13,93
0,3	14,41	14,89	15,37	15,85	16,33	16,81	17,29	17,77	18,25	18,73
0,4	19,21	19,69	20,17	20,65	21,13	21,61	22,10	22,58	23,06	23,54
0,5	24,02	24,50	24,98	25,42	25,94	26,42	26,90	27,38	27,86	28,34
0,6	28,82	29,30	29,78	30,22	30,74	31,22	31,70	32,18	32,66	33,14
0,7	33,62	34,10	34,58	35,06	35,54	36,02	36,51	36,99	37,47	37,95
0,8	38,43	38,91	39,39	39,87	40,35	40,83	41,31	41,79	42,27	42,75
0,9	43,23	43,71	44,19	44,67	45,15	45,63	46,11	46,59	47,07	47,55
1,0	48,03	48,51	48,99	49,47	49,95	50,43	50,91	51,40	51,88	52,36
1,1	52,84	53,32	53,80	54,28	54,76	55,24	55,72	56,20	56,68	57,16
1,2	57,64	58,12	58,60	59,08	59,56	60,04	60,52	61,00	61,48	61,96
1,3	62,44	62,92	63,40	63,88	64,36	64,84	65,32	65,81	66,29	66,77
1,4	67,25	67,73	68,21	68,69	69,17	69,65	70,13	70,61	71,09	71,57
1,5	72,05	72,53	73,01	73,49	73,97	74,45	74,93	75,41	75,89	76,37
1,6	76,85	77,33	77,81	78,29	78,77	79,25	79,73	80,22	80,70	81,18
1,7	81,66	82,14	82,62	83,10	83,58	84,06	84,54	85,02	85,50	85,98
1,8	86,46	86,94	87,42	87,90	88,38	88,86	89,34	89,82	90,30	90,78
1,9	91,26	91,74	92,22	92,70	93,18	93,66	94,14	94,63	95,11	95,59
2,0	96,07	96,55	97,03	97,51	97,99	98,47	98,95	99,43	99,91	100,39
2,1	100,87	101,35	101,83	102,31	102,79	103,27	103,75	104,23	104,71	105,19
2,2	105,67	106,15	106,63	107,11	107,59	108,07	108,55	109,03	109,52	110,00
2,3	110,48	110,96	111,44	111,92	112,40	112,88	113,36	113,84	114,32	114,80
2,4	115,28	115,76	116,24	116,72	117,20	117,68	118,16	118,64	119,12	119,60
2,5	120,08	120,56	121,04	121,52	122,00	122,48	122,96	123,44	123,93	124,41
2,6	124,89	125,37	125,85	126,33	126,81	127,29	127,77	128,25	128,73	129,21
2,7	129,69	130,17	130,65	131,13	131,61	132,09	132,57	133,05	133,53	134,01
2,8	134,49	134,97	135,45	135,93	136,41	136,89	137,37	137,85	138,34	138,82
2,9	139,30	139,78	140,26	140,74	141,22	141,70	142,18	142,66	143,14	143,62
3,0	144,10	144,58	145,06	145,54	146,02	146,50	146,98	147,46	147,94	148,42
3,1	148,90	149,38	149,96	150,44	150,82	151,30	151,78	152,26	152,74	153,23
3,2	153,71	154,19	154,67	155,15	155,63	156,11	156,59	157,07	157,55	158,03
3,3	158,51	158,99	159,47	159,95	160,43	160,91	161,39	161,87	162,35	162,83
3,4	163,31	163,79	164,27	164,75	165,23	165,71	166,19	166,67	167,15	167,64
3,5	168,12	168,60	169,08	169,56	170,04	170,52	171,00	171,48	171,96	172,44
3,6	172,92	173,40	173 88	174,36	174,84	175,32	175,80	176,28	176,76	177,24
3,7	177,72	178,20	178,68	179,16	179,64	180,12	180,60	181,08	181,56	182,05
3,8	182,52	183,01	183,49	183,97	184,45	184,93	185,41	185,89	186,37	186,85
3,9	187,33	187,81	188,29	188,77	189,25	189,73	190,21	190,69	191,17	191,65
4,0	192,13	192,61	193,09	193,57	194,05	194,53	195,01	195,49	195,97	196,45
4,1	196,94	197,42	197,90	198,38	198,86	199,34	199,82	200,30	200,78	201,26
4,2	201,74	202,22	202,70	203,18	203,66	204,14	204,62	205,10	205,58	206,06
4,3	206,54	207,02	207,50	207,98	208,46	208,94	209,42	209,90	210,38	210,86
4,4	211,35	211,83	212,31	212,79	213,27	213,75	214,23	214,71	215,19	215,67
4,5	216,15	216,63	217,11	217,59	218,07	218,55	219,03	219,51	219,99	220,47
4,6	220,95	221,43	221,91	222,39	222,87	223,35	223,83	224,31	224,79	225,27
4,7	225,76	226,24	226,72	227,20	227,68	228,16	228,64	229,12	229,60	230,08
4,8	230,56	231,04	231,52	232,00	232,48	232,96	233,44	233,92	234,40	234,88
4,9	235,36	235,84	236,32	236,80	237,28	237,76	238,24	238,72	239,20	239,68
5,0	240,17	240,65	241,13	241,61	242,09	242,57	243,05	243,53	244,01	244,49

Whole and tenths of meq	Hundredths of a meq									
	0	1	2	3	4	5	6	7	8	9
5,1	244,97	245,45	245,93	246,41	246,89	247,37	247,85	248,33	248,81	249,29
5,2	249,77	250,25	250,73	251,21	251,69	252,17	252,65	253,13	253,61	254,09
5,3	254,57	255,06	255,54	256,02	256,50	256,98	257,46	257,94	258,42	258,90
5,4	259,38	259,86	260,34	260,82	261,30	261,78	262,26	262,74	263,22	263,70
5,5	264,18	264,68	265,14	265,62	266,10	266,58	267,06	267,54	268,02	268,50
5,6	268,98	269,47	269,95	270,43	270,91	271,39	271,87	272,35	272,83	273,31
5,7	273,79	274,27	274,75	275,23	275,71	276,19	276,67	277,15	277,63	278,11
5,8	278,59	279,07	279,55	280,03	280,51	280,99	281,47	281,95	282,43	282,91
5,9	283,39	283,88	284,36	284,84	285,32	285,80	286,28	286,76	287,24	287,72
6,0	288,20	288,68	289,16	289,64	290,12	290,60	291,08	291,56	292,04	292,52
6,1	293,00	293,48	293,96	294,44	294,92	295,40	295,88	296,36	296,84	297,32
6,2	297,80	298,28	298,77	299,25	299,73	300,21	300,69	301,17	301,65	302,13
6,3	302,61	303,09	303,57	304,05	304,53	305,01	305,49	305,97	306,45	306,93
6,4	307,41	307,89	308,37	308,85	309,33	309,81	310,29	310,77	311,25	311,73
6,5	312,21	312,69	313,18	313,66	314,14	314,62	315,10	315,58	316,06	316,54
6,6	317,02	317,50	317,98	318,46	318,94	319,42	319,90	320,38	320,86	321,34
6,7	321,82	322,30	322,78	323,26	323,74	324,22	324,70	325,18	325,66	326,14
6,8	326,62	327,10	327,59	328,07	328,55	329,03	329,51	329,99	330,47	330,95
6,9	331,43	331,91	332,39	332,87	333,35	333,83	334,31	334,79	335,27	335,75
7,0	336,23	336,71	337,19	337,67	338,15	338,63	339,11	339,59	340,07	340,55
7,1	341,03	341,51	341,99	342,48	342,96	343,44	343,92	344,40	344,88	345,36
7,2	345,84	346,32	346,80	347,28	347,76	348,24	348,72	349,20	349,68	350,16
7,3	350,64	351,12	351,60	352,08	352,56	353,04	353,52	354,00	354,48	354,96
7,4	355,44	355,92	356,40	356,89	357,37	357,85	358,33	358,81	359,29	359,77
7,5	360,25	360,73	361,21	361,69	362,17	362,65	363,13	363,61	364,09	364,57
7,6	365,05	365,53	366,01	366,49	366,97	367,45	367,93	368,41	368,89	369,37
7,7	369,85	370,33	370,81	371,30	371,78	372,26	372,74	373,22	373,70	374,18
7,8	374,66	375,14	375,62	376,10	376,58	377,06	377,54	378,02	378,50	378,98
7,9	379,46	379,94	380,42	380,90	381,38	381,86	382,34	382,82	383,30	383,78
8,0	384,26	384,74	385,22	385,70	386,19	386,67	387,15	387,63	388,11	388,59
8,1	389,07	389,55	390,03	390,51	390,99	391,47	391,95	392,43	392,91	393,39
8,2	393,87	394,35	394,83	395,31	395,79	396,27	396,75	397,23	397,71	398,19
8,3	398,67	399,15	399,63	400,11	400,60	401,08	401,56	402,04	402,52	403,00
8,4	403,48	403,96	404,44	404,92	405,40	405,88	406,36	406,84	407,32	407,80
8,5	408,28	408,76	409,24	409,72	410,20	410,68	411,16	411,64	412,12	412,60
8,6	413,08	413,56	414,04	414,52	415,01	415,49	415,97	416,45	416,93	417,41
8,7	417,89	418,37	418,85	419,33	419,81	420,29	420,77	421,25	421,73	422,21
8,8	422,69	423,17	423,65	424,13	424,61	425,09	425,57	426,05	426,53	427,01
8,9	427,49	427,97	428,45	428,93	429,42	429,90	430,38	430,86	431,34	431,82
9,0	432,30	432,78	433,26	433,74	434,22	434,70	435,18	435,66	436,14	436,62
9,1	437,10	437,58	438,06	438,54	439,02	439,50	439,98	440,46	440,94	441,42
9,2	441,90	442,38	442,86	443,34	443,82	444,31	444,79	445,27	445,75	446,23
9,3	446,71	447,19	447,67	448,15	448,63	449,11	449,59	450,07	450,55	451,03
9,4	451,51	451,99	452,47	452,95	453,43	453,91	454,39	454,87	455,35	455,83
9,5	456,31	456,79	457,27	457,75	458,23	458,72	459,20	459,68	460,16	460,64
9,6	461,12	461,60	462,08	462,56	463,04	463,52	464,00	464,48	464,96	465,44
9,7	465,92	466,40	466,88	467,36	467,84	468,32	468,80	469,28	469,76	470,24
9,8	470,72	471,20	471,68	472,16	472,64	473,13	473,61	474,09	474,57	475,05
9,9	475,53	476,01	476,49	476,97	477,45	477,93	478,41	478,89	479,37	479,85
10,0	480,33	480,81	481,29	481,77	482,25	482,73	483,21	483,69	484,17	484,65

meq	100	200	300	400	500	600	700	800	900	1000
mg	4803,3	9606,6	14409,9	19213,2	24016,5	28819,8	33623,1	38426,4	43229,7	48033

Table 32

Conversion of Milligram—Equivalents of HCO_3^- into Milligrams
(equivalent weight of HCO_3^- = 61.019)

Whole and tenths of meq	Hundredths of a meq									
	0	1	2	3	4	5	6	7	8	9
0,0	—	0,6	1,2	1,8	2,4	3,1	3,7	4,3	4,9	5,5
0,1	6,10	6,71	7,32	7,93	8,54	9,15	9,76	10 37	10,98	11,59
0,2	12,20	12,81	13,42	14.03	14,64	15,25	15,86	16,48	17,09	17,70
0,3	18,31	18,92	19,53	20,14	20,75	21,36	21,97	22,58	23,19	23,80
0,4	24,41	25,02	25,63	26,24	26,85	27,46	28,07	28,68	29,29	29,90
0,5	30,51	31,12	31,73	32,34	32,95	33,56	34,17	34,78	35,39	36,00
0,6	36,61	37,22	37,83	38,44	39,05	39,66	40,27	40,88	41,49	42,10
0,7	42,71	43,32	43.93	44,54	45,15	45,76	46,37	46,98	47,59	48,21
0,8	48,82	49,43	50,04	50,65	51,26	51,87	52,48	53,09	53,70	54,31
0,9	54,92	55,53	56,14	56,75	57,36	57,97	58,58	59,19	59,80	60,41
1,0	61,02	61,63	62,24	62,85	63,46	64,07	64,68	65,29	65,90	66,51
1,1	67,12	67,73	68,34	68,95	69,56	70,17	70,78	71,39	72,00	72,61
1,2	73,22	73,83	74,44	75,05	75,66	76,27	76,88	77,49	78,10	78,71
1,3	79,32	79,93	80,55	81,16	81,77	82,38	82,99	83,60	84,21	84,82
1,4	85,43	86,04	86,65	87,26	87,87	88,48	89,09	89,70	90,31	90,92
1,5	91,53	92,14	92,75	93,36	93,97	94,58	95,19	95,80	96,41	97,02
1,6	97,63	98,24	98,85	99,46	100,07	100,68	101,29	101,90	102,51	103,12
1,7	103,73	104,34	104,95	105,56	106,17	106,78	107,39	108,00	108,61	109,22
1,8	109,83	110,44	111,05	111,66	112,27	112,89	113,50	114,11	114,72	115,33
1,9	115,94	116,55	117,16	117,77	118,38	118,99	119,60	120,21	120,82	121,43
2,0	122,04	122,65	123,26	123,87	124,48	125,09	125,70	126,31	126,92	127,53
2,1	128,14	128,75	129,36	129,97	130,58	131,19	131,80	132,41	133,02	133,63
2,2	134,24	134,85	135,46	136,07	136,68	137,29	137,90	138,51	139,12	139,73
2,3	140,34	140,95	141,56	142,17	142,78	143,39	144,00	144,62	145,23	145,84
2,4	146,45	147,06	147,67	148,28	148,89	149,50	150,11	150,72	151,33	151,94
2,5	152,55	153,16	153,77	154,38	154,99	155,60	156,21	156,82	157,43	158,04
2,6	158,65	159,26	159,87	160,48	161,09	161,70	162,31	162,92	163,53	164,14
2,7	164,75	165,36	165,97	166,58	167,19	167,80	168,41	169,02	169,63	170,24
2,8	170,85	171,46	172,07	172,68	173,29	173,90	174,51	175,12	175,73	176,34
2,9	176,96	177,57	178,18	178,79	179,40	180,01	180,62	181,23	181,84	182,45
3,0	183,06	183,67	184,28	184,89	185,50	186,11	186,72	187,33	187,94	188,55
3,1	189,16	189,77	190,38	190,99	191,60	192,21	192,82	193,43	194,04	194,65
3,2	195,26	195,87	196,48	197,09	197,70	198,31	198,92	199,53	200,14	200,75
3,3	201,36	201,97	202,58	203,19	203,80	204,41	205,02	205,63	206,24	206,85
3,4	207,46	208,07	208,68	209,30	209,91	210,52	211,13	211,74	212,35	212,96
3,5	213,57	214,18	214,79	215,40	216,01	216,62	217,23	217,84	218,45	219,06
3,6	219,67	220,28	220,89	221,50	222,11	222,72	223,33	223,94	224,55	225,16
3,7	225,77	226,38	226,99	227,60	228,21	228,82	229,43	230,04	230,65	231,26
3,8	231,87	232,48	233,09	233,70	234,31	234,92	235,53	236,14	236,75	237,36
3,9	237,97	238,58	239,19	239,80	240,41	241,03	241,64	242,25	242,86	243,47
4,0	244,08	244,69	245,30	245,91	246,52	247,13	247,74	248,35	248,96	249,57
4,1	250,18	250,79	251,40	252,01	252,62	253,23	253,84	254,45	255 06	255,67
4,2	256,28	256,89	257,50	258,11	258,72	259,33	259,94	260,55	261,16	261,77
4,3	262,38	262,99	263,60	264,21	264,82	265,43	266,04	266,65	267,26	267,87
4,4	268,48	269,09	269,70	270,31	270,92	271,53	272,14	272,75	273,37	273,98
4,5	274,59	275,20	275,81	276,42	277,03	277,64	278,25	278,86	279 47	280,08
4,6	280,69	281,30	281,91	282,52	283,13	283,74	284,35	284,96	285,57	286,18
4,7	286,79	287,40	288,01	288,62	289,23	289,84	290,45	291,06	291,67	292,28
4,8	292,89	293,50	294,11	294,72	295,33	295,94	296,55	297,16	297,77	298,38
4,9	298,99	299,60	300,21	300,82	301,43	302,04	302,65	303,26	303,87	304,48
5,0	305,09	305,71	306,32	306,93	307,54	308,15	308,76	309,37	309,98	310,59

Whole and tenths of meq	Hundredths of a meq									
	0	1	2	3	4	5	6	7	8	9
5,1	311,20	311,81	312,42	313,03	313,64	314,25	314,86	315,47	316,08	316,69
5,2	317,30	317,91	318,52	319,13	319,74	320,35	320,96	321,57	322,18	322,79
5,3	323,40	324,01	324,62	325,23	325,84	326,45	327,06	327,67	328,28	328,89
5,4	329,50	330,11	330,72	331,33	331,94	332,55	333,16	333,77	334,38	334,99
5,5	335,60	336,21	336,82	337,44	338,05	338,66	339,27	339,88	340,49	341,10
5,6	341,71	342,32	342,93	343,54	344,15	344,76	345,37	345,98	346,59	347,20
5,7	347,81	348,42	349,03	349,64	350,25	350,86	351,47	352,08	352,69	353,30
5,8	353,91	354,52	355,13	355,74	356,35	356,96	357,57	358,18	358,79	359,40
5,9	360,01	360,62	361,23	361,84	362,45	363,06	363,67	364,28	364,89	365,50
6,0	366,11	366,72	367,33	367,94	368,55	369,16	369,78	370,39	371,00	371,61
6,1	372,22	372,83	373,44	374,05	374,66	375,27	375,88	376,49	377,10	377,71
6,2	378,32	378,93	379,54	380,15	380,76	381,37	381,98	382,59	383,20	383,81
6,3	384,42	385,03	385,64	386,25	386,86	387,47	388,08	388,69	389,30	389,91
6,4	390,52	391,13	391,74	392,35	392,96	393,57	394,18	394,79	395,40	396,01
6,5	396,62	397,23	397,84	398,45	399,06	399,67	400,28	400,89	401,51	402,12
6,6	402,78	403,34	403,95	404,56	405,17	405,78	406,39	407,00	407,61	408,22
6,7	408,83	409,44	410,05	410,66	411,27	411,88	412,49	413,10	413,71	414,32
6,8	414,93	415,54	416,15	416,76	417,37	417,98	418,59	419,20	419,81	420,42
6,9	421,03	421,64	422,25	422,86	423,47	424,08	424,69	425,30	425,91	426,52
7,0	427,13	427,74	428,35	428,96	429,57	430,18	430,79	431,40	432,01	432,62
7,1	433,23	433,85	434,46	435,07	435,68	436,29	436,90	437,51	438,12	438,73
7,2	439,34	439,95	440,56	441,17	441,78	442,39	443,00	443,61	444,22	444,83
7,3	445,44	446,05	446,66	447,27	447,88	448,49	449,10	449,71	450,32	450,93
7,4	451,54	452,15	452,76	453,37	453,98	454,59	455,20	455,81	456,42	457,03
7,5	457,64	458,25	458,86	459,47	460,08	460,69	461,30	461,91	462,52	463,13
7,6	463,74	464,35	464,96	465,57	466,19	466,80	467,41	468,02	468,63	469,24
7,7	469,85	470,46	471,07	471,68	472,29	472,90	473,51	474,12	474,73	475,34
7,8	475,95	476,56	477,17	477,78	478,39	479,00	479,61	480,22	480,83	481,44
7,9	482,05	482,66	483,27	483,88	484,49	485,10	485,71	486,32	486,93	487,54
8,0	488,15	488,76	489,37	489,98	490,59	491,20	491,81	492,42	493,03	493,64
8,1	494,25	494,86	495,47	496,08	496,69	497,30	497,92	498,53	499,14	499,75
8,2	500,36	500,97	501,58	502,19	502,80	503,41	504,02	504,63	505,24	505,85
8,3	506,46	507,07	507,68	508,29	508,90	509,51	510,12	510,73	511,34	511,95
8,4	512,56	513,17	513,78	514,39	515,00	515,61	516,22	516,83	517,44	518,05
8,5	518,66	519,27	519,88	520,49	521,10	521,71	522,32	522,93	523,54	524,15
8,6	524,76	525,37	525,98	526,59	527,20	527,81	528,42	529,03	529,64	530,26
8,7	530,87	531,48	532,09	532,70	533,31	533,92	534,53	535,14	535,75	536,36
8,8	536,97	537,58	538,19	538,80	539,41	540,02	540,63	541,24	541,85	542,46
8,9	543,07	543,68	544,29	544,90	545,51	546,12	546,73	547,34	547,95	548,56
9,0	549,17	549,78	550,39	551,00	551,61	552,22	552,83	553,44	554,05	554,66
9,1	555,27	555,88	556,49	557,10	557,71	558,32	558,93	559,54	560,15	560,76
9,2	561,37	561,98	562,60	563,21	563,82	564,43	565,04	565,65	566,26	566,87
9,3	567,48	568,09	568,70	569,31	569,92	570,53	571,14	571,75	572,36	572,97
9,4	573,58	574,19	574,80	575,41	576,02	576,63	577,24	577,85	578,46	579,07
9,5	579,68	580,29	580,90	581,51	582,12	582,73	583,34	583,95	584,56	585,17
9,6	585,78	586,39	587,00	587,61	588,22	588,83	589,44	590,05	590,66	591,27
9,7	591,88	592,49	593,10	593,71	594,33	594,94	595,55	596,16	596,77	597,38
9,8	597,99	598,60	599,21	599,82	600,43	601,04	601,65	602,26	602,87	603,48
9,9	604,09	604,70	605,31	605,92	606,53	607,14	607,75	608,36	608,97	609,58
10,0	610,19	610,80	611,41	612,02	612,63	613,24	613,85	614,46	615,07	615,68

Table 33

Conversion of Milligram—Equivalents of CO_3^{2-} into Milligrams
(equivalent weight of CO_3^{2-} = 30.0055)

Whole and tenths of meq	Hundredths of a meq									
	0	1	2	3	4	5	6	7	8	9
0,0	—	0,3	0,6	0,9	1,2	1,5	1,8	2,1	2,4	2,7
0,1	3,00	3,30	3,60	3,90	4,20	4,50	4,80	5,10	5,40	5,70
0,2	6,00	6,30	6,60	6,90	7,20	7,50	7,80	8,10	8,40	8,70
0,3	9,00	9,30	9,60	9,90	10,20	10,50	10,80	11,10	11,40	11,70
0,4	12,00	12,30	12,60	12,90	13,20	13,50	13,80	14,10	14,40	14,70
0,5	15,00	15,30	15,60	15,90	16,20	16,50	16,80	17,10	17,40	17,70
0,6	18,00	18,30	18,60	18,90	19,20	19,50	19,80	20,10	20,40	20,70
0,7	21,00	21,30	21,60	21,90	22,20	22,50	22,80	23,10	23,40	23,70
0,8	24,00	24,30	24,60	24,90	25,20	25,50	25,80	26,10	26,40	26,70
0,9	27,00	27,31	27,61	27,91	28,21	28,51	28,81	29,11	29,41	29,71
1,0	30,01	30,31	30,61	30,91	31,21	31,51	31,81	32,11	32,41	32,71
1,1	33,01	33,31	33,61	33,91	34,21	34,51	34,81	35,11	35,41	35,71
1,2	36,01	36,31	36,61	36,91	37,21	37,51	37,81	38,11	38,41	38,71
1,3	39,01	39,31	39,61	39,91	40,21	40,51	40,81	41,11	41,41	41,71
1,4	42,01	42,31	42,61	42,91	43,21	43,51	43,81	44,11	44,41	44,71
1,5	45,01	45,31	45,61	45,91	46,21	46,51	46,81	47,11	47,41	47,71
1,6	48,01	48,31	48,61	48,91	49,21	49,51	49,81	50,11	50,41	50,71
1,7	51,01	51,31	51,61	51,91	52,21	52,51	52,81	53,11	53,41	53,71
1,8	54,01	54,31	54,61	54,91	55,21	55,51	55,81	56,11	56,41	56,71
1,9	57,01	57,31	57,61	57,91	58,21	58,51	58,81	59,11	59,41	59,71
2,0	60,01	60,31	60,61	60,91	61,21	61,51	61,81	62,11	62,41	62,71
2,1	63,01	63,31	63,61	63,91	64,21	64,51	64,81	65,11	65,41	65,71
2,2	66,01	66,31	66,61	66,91	67,21	67,51	67,81	68,11	68,41	68,71
2,3	69,01	69,31	69,61	69,91	70,21	70,51	70,81	71,11	71,41	71,71
2,4	72,01	72,31	72,61	72,91	73,21	73,51	73,81	74,11	74,41	74,71
2,5	75,01	75,31	75,61	75,91	76,21	76,51	76,81	77,11	77,41	77,71
2,6	78,01	78,31	78,61	78,91	79,21	79,51	79,81	80,11	80,41	80,71
2,7	81,01	81,31	81,61	81,92	82,22	82,52	82,82	83,12	83,42	83,72
2,8	84,02	84,32	84,62	84,92	85,22	85,52	85,82	86,12	86,42	86,72
2,9	87,02	87,32	87,62	87,92	88,22	88,52	88,82	89,12	89,42	89,72
3,0	90,02	90,32	90,62	90,92	91,22	91,52	91,82	92,12	92,42	92,72
3,1	93,02	93,32	93,62	93,92	94,22	94,52	94,82	95,12	95,42	95,72
3,2	96,02	96,32	96,62	96,92	97,22	97,52	97,82	98,12	98,42	98,72
3,3	99,02	99,32	99,62	99,92	100,22	100,52	100,82	101,12	101,42	101,72
3,4	102,02	102,32	102,62	102,92	103,22	103,52	103,82	104,12	104,42	104,72
3,5	105,02	105,32	105,62	105,92	106,22	106,52	106,82	107,12	107,42	107,72
3,6	108,02	108,32	108,62	108,92	109,22	109,52	109,82	110,12	110,42	110,72
3,7	111,02	111,32	111,62	111,92	112,22	112,52	112,82	113,12	113,42	113,72
3,8	114,02	114,32	114,62	114,92	115,22	115,52	115,82	116,12	116,42	116,72
3,9	117,02	117,32	117,62	117,92	118,22	118,52	118,82	119,12	119,42	119,72
4,0	120,02	120,32	120,62	120,92	121,22	121,52	121,82	122,12	122,42	122,72
4,1	123,02	123,32	123,62	123,92	124,22	124,52	124,82	125,12	125,42	125,72
4,2	126,02	126,32	126,62	126,92	127,22	127,52	127,82	128,12	128,42	128,72
4,3	129,02	129,32	129,62	129,92	130,22	130,52	130,82	131,12	131,42	131,72
4,4	132,02	132,32	132,62	132,92	133,22	133,52	133,82	134,12	134,42	134,72
4,5	135,02	135,32	135,62	135,92	136,22	136,53	136,83	137,13	137,43	137,73
4,6	138,03	138,33	138,63	138,93	139,23	139,53	139,83	140,13	140,43	140,73
4,7	141,03	141,33	141,63	141,93	142,23	142,53	142,83	143,13	143,43	143,73
4,8	144,03	144,33	144,63	144,93	145,23	145,53	145,83	146,13	146,43	146,73
4,9	147,03	147,33	147,63	147,93	148,23	148,53	148,83	149,13	149,43	149,73
5,0	150,03	150,33	150,63	150,93	151,23	151,53	151,83	152,13	152,43	152,73

Table 34

Conversion of Milligram—Equivalents of NO_3^- into Milligrams
(equivalent weight of NO_3^- = 62.008)

Whole and tenths of meq	Hundredths of a meq									
	0	1	2	3	4	5	6	7	8	9
0,0	—	0,6	1,2	1,9	2,5	3,1	3,7	4,3	5,0	5,6
0,1	6,20	6,82	7,44	8,06	8,68	9,30	9,92	10,54	11,16	11,78
0,2	12,40	13,02	13,64	14,26	14,88	15,50	16,12	16,74	17,36	17,98
0,3	18,60	19,22	19,84	20,46	21,08	21,70	22,32	22,94	23,56	24,18
0,4	24,80	25,42	26,04	26,66	27,28	27,90	28,52	29,14	29,76	30,38
0,5	31,00	31,62	32,24	32,86	33,48	34,10	34,72	35,34	35,96	36,58
0,6	37,20	37,82	38,44	39,07	39,69	40,31	40,93	41,55	42,17	42,79
0,7	43,41	44,03	44,65	45,27	45,89	46,51	47,13	47,75	48,37	48,99
0,8	49,61	50,23	50,85	51,47	52,09	52,71	53,33	53,95	54,57	55,19
0,9	55,81	56,43	57,05	57,67	58,29	58,91	59,53	60,15	60,77	61,39
1,0	62,01	62,63	63,25	63,87	64,49	65,11	65 73	66,35	66,97	67,59
1,1	68,21	68,83	69,45	70,07	70,69	71,31	71,93	72,55	73,17	73.79
1,2	74,41	75,03	75,65	76,27	76,89	77,51	78,13	78,75	79,37	79,99
1,3	80,61	81,23	81,85	82,47	83,09	83,71	84,33	84,95	85,57	86,19
1,4	86,81	87,43	88,05	88,67	89,29	89,91	90,53	91,15	91,77	92,39
1,5	93,01	93,63	94,25	94,87	95,49	96,11	96,73	97,35	97,97	93,59
1,6	99,21	99,83	100,45	101,07	101,69	102,31	102,93	103,55	104,17	104,79
1,7	105,41	106,03	106,65	107,27	107,89	108,51	109,13	109,75	110,37	110,99
1,8	111,61	112,23	112,85	113,47	114,09	114,71	115,33	115,95	116,58	117,20
1,9	117,82	118,44	119,06	119,68	120,30	120,92	121,54	122,16	122,78	123,40
2,0	124,02	124,64	125,26	125,88	126,50	127,12	127,74	128,36	129,98	129,60
2,1	130,22	130,84	131,46	132,08	132,70	133,32	133,94	134,56	135,18	135,80
2,2	136,42	137,04	137,66	138,28	138,90	139,52	140,14	140,76	141,38	142,00
2,3	142,62	143,24	143,86	144,48	145,10	145,72	146,34	146,96	147,58	148,20
2,4	148,82	149,44	150,06	150,68	151,30	151,92	152,54	153,16	153,78	154,40
2,5	155,02	155,64	156,26	156 88	157,50	158,12	158,74	159,36	159,98	160,60
2,6	161,22	161,84	162,46	163,08	163,70	164,32	164,94	165,56	166,18	166,80
2,7	167,42	168,04	168,66	169,28	169,90	170,52	171,14	171,76	172,38	173.00
2,8	173,62	174,24	174,86	175,48	176,10	176,72	177,34	177,96	178,58	179,20
2,9	179,82	180,44	181,06	181,68	182,30	182,92	183,54	184,16	184,78	185,40
3,0	186,02	186,64	187,26	187,88	188,50	189,12	189,74	190,36	190,98	191,60
3,1	192,22	192,84	193,46	194,09	194,71	195,33	195,95	196,57	197,19	197,81
3,2	198,43	199,05	199,67	200,29	200,91	201,53	202,15	202 71	203,39	204,01
3,3	204,63	205,25	205,87	206,49	207,11	207,73	208,35	208,97	209,59	210,21
3,4	210,83	211,45	212,07	212,69	213,31	213,93	214,55	215,17	215,79	216,41
3,5	217,03	217,65	218,27	218,89	219,51	220,13	220,75	221,37	221,99	222,61
3,6	223,23	223,85	224,47	225,09	225,71	226,33	226,95	227,57	228,19	228,81
3,7	229,43	230,05	230,67	231,29	231,91	232,53	233,15	233,77	234,39	235,01
3,8	235,63	236,25	236,87	237,49	238,11	238,73	239,35	239,97	240,59	241,21
3,9	241,83	242,45	243,07	243,69	244,31	244,93	245,55	246,17	246,79	247,41
4,0	248,03	248,65	249,27	249,89	250,51	251,13	251,75	252,37	252,99	253,61
4,1	254,23	254,85	255,47	256,09	256,71	257,33	257,95	258,57	259,19	259,81
4,2	260,43	261,05	261,67	262,29	262,91	263,53	264,15	264,77	265,39	266,01
4,3	266,63	267,25	267,87	268,49	269,11	269,73	270,35	270,97	271,60	272,22
4,4	272,84	273,46	274,08	274,70	275,32	275,94	276,56	277,18	277,80	278,42
4,5	279,04	279,66	280,28	280,90	281,52	282,14	282,76	283,38	284,00	284,62
4,6	285,24	285,86	286,48	287,10	287,72	288,34	288,96	289,58	290,20	290,82
4,7	291,44	292,06	292,68	293,30	293,92	294,54	295,16	295,78	296,40	297,02
4,8	297,64	298,26	298,88	299,50	300,12	300,74	301,36	301,98	302,60	303,22
4,9	303,84	304,46	305,08	305,70	306,32	306,94	307,56	308,18	308,80	309,42
5,0	310,04	310,66	311,28	311,90	312,52	313,14	313,76	314,38	315,00	315,62

Table 35

Conversion of Milligram—Equivalents of NO_2^- into Milligrams
(equivalent weight of NO_2^- = 46.008)

Whole and tenths of meq	Hundredths of a meq									
	0	1	2	3	4	5	6	7	8	9
0,0	—	0,5	0,9	1,4	1,8	2,3	2,8	3,2	3,7	4,1
0,1	4,60	5,06	5,52	5,98	6,44	6,90	7,36	7,82	8,28	8,74
0,2	9,20	9,66	10,12	10,58	11,04	11,50	11,96	12,42	12,88	13,34
0,3	13,80	14,26	14,72	15,18	15,64	16,10	16,56	17,02	17,48	17,94
0,4	18,40	18,86	19,32	19,78	20,24	20,70	21,16	21,62	22,08	22,54
0,5	23,00	23,46	23,92	24,38	24,84	25,30	25,76	26,22	26,68	27,14
0,6	27,60	28,06	28,52	28,98	29,45	29,91	30,37	30,83	31,29	31,75
0,7	32,21	32,67	33,13	33,59	34,05	34,51	34,97	35,43	35,89	36,35
0,8	36,81	37,27	37,73	38,19	38,65	39,11	39,57	40,03	40,49	40,95
0,9	41,41	41,87	42,33	42,79	43,25	43,71	44,17	44,63	45,09	45,55
1,0	46,01	46,47	46,93	47,39	47,85	48,31	48,77	49,23	49,69	50,15

Table 36

Conversion of Milligram—Equivalents of Br^- into Milligrams
(equivalent weight of Br^- = 79.916)

Whole and tenths of meq	Hundredths of a meq									
	0	1	2	3	4	5	6	7	8	9
0,0	—	0,8	1,6	2,4	3,2	4,0	4,8	5,6	6,4	7,2
0,1	7,99	8,79	9,59	10,39	11,19	11,99	12,79	13,59	14,38	15,18
0,2	15,98	16,78	17,58	18,38	19,18	19,98	20,79	21,58	22,38	23,18
0,3	23,97	24,77	25,57	26,37	27,17	27,97	28,77	29,57	30,37	31,17
0,4	31,97	32,77	33,56	34,36	35,16	35,96	36,76	37,56	38,36	39,16
0,5	39,96	40,76	41,56	42,36	43,15	43,95	44,75	45,55	46,35	47,15
0,6	47,95	48,75	49,55	50,35	51,15	51,95	52,74	53,54	54,34	55,14
0,7	55,94	56,74	57,54	58,34	59,14	59,94	60,74	61,54	62,33	63,13
0,8	63,93	64,73	65,53	66,33	67,13	67,93	68,73	69,53	70,33	71,13
0,9	71,92	72,72	73,52	74,32	75,12	75,92	76,72	77,52	78,32	79,12
1,0	79,92	80,72	81,51	82,31	83,11	83,91	84,71	85,51	86,31	87,11

Table 37

Conversion of Milligram—Equivalents of I⁻ into Milligrams
(equivalent weight of I⁻ = 126.91)

Whole and tenths of meq	Hundredths of a meq									
	0	1	2	3	4	5	6	7	8	9
0,0	—	1,3	2,5	3,8	5,1	6,3	7,6	8,9	10,1	11,4
0,1	12,69	13,96	15,23	16,50	17,77	19,04	20,31	21,57	22,84	24,11
0,2	25,38	26,65	27,92	29,19	30,46	31,73	33,00	34,27	35,53	36,80
0,3	38,07	39,34	40,61	41,88	43,15	44,42	45,69	46,96	48,23	49,49
0,4	50,76	52,03	53,30	54,57	55,84	57,11	58,38	59,65	60,92	62,19
0,5	63,46	64,72	65,99	67,26	68,53	69,80	71,07	72,34	73,61	74,88
0,6	76,15	77,42	78,68	79,95	81,22	82,49	83,76	85,03	86,30	87,57
0,7	88,84	90,11	91,38	92,64	93,91	95,18	96,45	97,72	98,99	100,26
0,8	101,53	102,80	104,07	105,34	106,60	107,87	109,14	110,41	111,68	112,95
0,9	114,22	115,49	116,76	118,03	119,30	120,56	121,83	123,10	124,37	125,64
1,0	126,91	128,18	129,45	130,72	131,99	133,26	134,52	135,79	137,06	138,33

Table 38

Conversion of Milligram—Equivalents of F⁻ into Milligrams
(equivalent weight of F⁻ = 19.00)

Whole and tenths of meq	Hundredths of a meq									
	0	1	2	3	4	5	6	7	8	9
0,0	—	0,2	0,4	0,6	0,8	0,9	1,1	1,3	1,5	1,7
0,1	1,90	2,09	2,28	2,47	2,66	2,85	3,04	3,23	3,42	3,61
0,2	3,80	3,99	4,18	4,37	4,56	4,75	4,94	5,13	5,32	5,51
0,3	5,70	5,89	6,08	6,27	6,46	6,65	6,84	7,03	7,22	7,41
0,4	7,60	7,79	7,98	8,17	8,36	8,55	8,74	8,93	9,12	9,31
0,5	9,50	9,69	9,88	10,07	10,26	10,45	10,64	10,83	11,02	11,21
0,6	11,40	11,59	11,78	11,97	12,16	12,35	12,54	12,73	12,92	13,11
0,7	13,30	13,49	13,68	13,87	14,06	14,25	14,44	14,63	14,82	15,01
0,8	15,20	15,39	15,58	15,77	15,96	16,15	16,34	16,53	16,72	16,91
0,9	17,10	17,29	17,48	17,67	17,86	18,05	18,24	18,43	18,62	18,81
1,0	19,00	19,19	19,38	19,57	19,76	19,95	20,14	20,33	20,52	20,71

49

Table 39

Conversion of Milligrams of Na_2O into Milligrams of Na^+

Whole mg of Na_2O	Tenths of a mg of Na_2O									
	0	1	2	3	4	5	6	7	8	9
0	0,74	0,1	0,1	0,2	0,3	0,4	0,4	0,5	0,6	0,7
1	0,74	0,82	0,89	0,96	1,04	1,11	1,19	1,26	1,34	1,41
2	1,48	1,56	1,63	1,71	1,78	1,85	1,93	2,00	2,08	2,15
3	2,23	2,30	2,37	2,45	2,52	2,60	2,67	2,74	2,82	2,89
4	2,97	3,04	3,12	3,19	3,26	3,34	3,41	3,49	3,56	3,63
5	3,71	3,78	3,86	3,93	4,01	4,08	4,15	4,23	4,30	4,38
6	4,45	4,53	4,60	4,67	4,75	4,82	4,90	4,97	5,04	5,12
7	5,19	5,27	5,34	5,42	5,49	5,56	5,64	5,71	5,79	5,86
8	5,93	6,01	6,08	6,16	6,23	6,31	6,38	6,45	6,53	6,60
9	6,68	6,75	6,83	6,90	6,97	7,05	7,12	7,20	7,27	7,34
10	7,42	7,49	7,57	7,64	7,72	7,79	7,86	7,94	8,01	8,09
11	8,16	8,23	8,31	8,38	8,46	8,53	8,61	8,68	8,75	8,83
12	8,90	8,98	9,05	9,12	9,20	9,27	9,35	9,42	9,50	9,57
13	9,64	9,72	9,79	9,87	9,94	10,02	10,09	10,16	10,24	10,31
14	10,39	10,46	10,53	10,61	10,68	10,76	10,83	10,91	10,98	11,05
15	11,13	11,20	11,28	11,35	11,42	11,50	11,57	11,65	11,72	11,80
16	11,87	11,94	12,02	12,09	12,17	12,24	12,31	12,39	12,46	12,54
17	12,61	12,69	12,76	12,83	12,91	12,98	13,06	13,13	13,21	13,28
18	13,35	13,43	13,50	13,58	13,65	13,72	13,80	13,87	13,95	14,02
19	14,10	14,17	14,24	14,32	14,39	14,47	14,54	14,61	14,69	14,76
20	14,84	14,91	14,99	15,06	15,13	15,21	15,28	15,36	15,43	15,50
21	15,58	15,65	15,73	15,80	15,88	15,95	16,02	16,10	16,17	16,25
22	16,32	16,40	16,47	16,54	16,62	16,69	16,77	16,84	16,91	16,99
23	17,06	17,14	17,21	17,29	17,36	17,43	17,51	17,58	17,66	17,73
24	17,80	17,88	17,95	18,03	18,10	18,18	18,25	18,32	18,40	18,47
25	18,55	18,62	18,69	18,77	18,84	18,92	18,99	19,07	19,14	19,21
26	19,29	19,36	19,44	19,51	19,59	19,66	19,73	19,81	19,88	19,96
27	20,03	20,10	20,18	20,25	20,33	20,40	20,48	20,55	20,62	20,70
28	20,77	20,85	20,92	20,99	21,07	21,14	21,22	21,29	21,37	21,44
29	21,52	21,59	21,66	21,74	21,81	21,88	21,96	22,03	22,11	22,18
30	22,26	22,33	22,40	22,48	22,55	22,63	22,70	22,78	22,85	22,92
31	23,00	23,07	23,15	23,22	23,29	23,37	23,44	23,52	23,59	23,67
32	23,74	23,81	23,89	23,96	24,04	24,11	24,18	24,26	24,33	24,41
33	24,48	24,56	24,63	24,70	24,78	24,85	24,93	25,00	25,07	25,15
34	25,22	25,30	25,37	25,45	25,52	25,59	25,67	25,74	25,82	25,89
35	25,97	26,04	26,11	26,19	26,26	26,34	26,41	26,48	26,56	26,63
36	26,71	26,78	26,86	26,93	27,00	27,08	27,15	27,23	27,30	27,37
37	27,45	27,52	27,60	27,67	27,75	27,82	27,89	27,97	28,04	28,12
38	28,19	28,26	28,34	28,41	28,49	28,56	28,64	28,71	28,78	28,86
39	28,93	29,01	29,08	29,16	29,23	29,30	29,38	29,45	29,53	29,60
40	29,67	29,75	29,82	29,90	29,97	30,05	30,12	30,19	30,27	30,34
41	30,42	30,49	30,56	30,64	30,71	30,79	30,86	30,94	31,01	31,08
42	31,16	31,23	31,31	31,38	31,45	31,53	31,60	31,68	31,75	31,83
43	31,90	31,97	32,05	32,12	32,20	32,27	32,35	32,42	32,49	32,57
44	32,64	32,72	32,79	32,86	32,94	33,01	33,09	33,16	33,24	33,31
45	33,38	33,46	33,53	33,61	33,68	33,75	33,83	33,90	33,98	34,05
46	34,13	34,20	34,27	34,35	34,42	34,50	34,57	34,64	34,72	34,79
47	34,87	34,94	35,02	35,09	35,16	35,24	35,31	35,39	35,46	35,54
48	35,61	35,68	35,76	35,83	35,91	35,98	36,05	36,13	36,20	36,28
49	36,35	36,43	36,50	36,57	36,65	36,72	36,80	36,87	36,94	37,02
50	37,09	37,17	37,24	37,32	37,39	37,46	37,54	37,61	37,69	37,76

Whole mg of Na₂O	Tenths of a mg of Na₂O									
	0	1	2	3	4	5	6	7	8	9
51	37,83	37,91	37,98	38,06	38,13	38,21	38,28	38,35	38,43	38,50
52	38,58	38,65	38,73	38,80	38,87	38,95	39,02	39,10	39,17	39,24
53	39,32	39,39	39,47	39,54	39,62	39,69	39,76	39,84	39,91	39,99
54	40,06	40,13	40,21	40,28	40,36	40,43	40,51	40,58	40,65	40,73
55	40,80	40,88	40,95	41,02	41,10	41,17	41,25	41,32	41,40	41,47
56	41,54	41,62	41,69	41,77	41,84	41,92	41,99	42,06	42,14	42,21
57	42,29	42,36	42,43	42,51	42,58	42,66	42,73	42,81	42,88	42,95
58	43,03	43,10	43,18	43,25	43,32	43,40	43,48	43,55	43,62	43,70
59	43,77	43,84	43,92	43,99	44,07	44,14	44,21	44,29	44,36	44,44
60	44,51	44,59	44,66	44,73	44,81	44,88	44,96	45,03	45,11	45,18
61	45,25	45,33	45,40	45,48	45,55	45,62	45,70	45,77	45,85	45,92
62	46,00	46,07	46,14	46,22	46,29	46,37	46,44	46,51	46,59	46,66
63	46,74	46,81	46,89	46,96	47,03	47,11	47,18	47,26	47,33	47,40
64	47,48	47,55	47,63	47,70	47,78	47,85	47,92	48,00	48,07	48,15
65	48,22	48,30	48,37	48,44	48,52	48,59	48,67	48,74	48,81	48,89
66	48,96	49,04	49,11	49,19	49,26	49,33	49,41	49,48	49,56	49,63
67	49,70	49,78	49,85	49,93	50,00	50,08	50,15	50,22	50,30	50,37
68	50,45	50,52	50,59	50,67	50,74	50,82	50,89	50,97	51,04	51,11
69	51,19	51,26	51,34	51,41	51,49	51,56	51,63	51,71	51,78	51,86
70	51,93	52,00	52,08	52,15	52,23	52,30	52,38	52,45	52,52	52,60
71	52,67	52,75	52,82	52,89	52,97	53,04	53,12	53,19	53,27	53,34
72	53,41	53,49	53,56	53,64	53,71	53,78	53,86	53,93	54,01	54,08
73	54,16	54,23	54,30	54,38	54,46	54,53	54,60	54,68	54,75	54,82
74	54,90	54,97	55,05	55,12	55,19	55,27	55,34	55,42	55,49	55,57
75	55,64	55,71	55,79	55,86	55,94	56,01	56,08	56,16	56,23	56,31
76	56,38	56,46	56,53	56,60	56,68	56,75	56,83	56,90	56,97	57,05
77	57,12	57,20	57,27	57,35	57,42	57,49	57,57	57,64	57,72	57,79
78	57,87	57,94	58,01	58,09	58,16	58,24	58,31	58,38	58,46	58,53
79	58,61	58,68	58,76	58,83	58,90	58,98	59,05	59,13	59,20	59,27
80	59,35	59,42	59,50	59,57	59,65	59,72	59,79	59,87	59,94	60,02
81	60,09	60,16	60,24	60,31	60,39	60,46	60,54	60,61	60,68	60,76
82	60,83	60,91	60,98	61,06	61,13	61,20	61,28	61,35	61,43	61,50
83	61,57	61,65	61,72	61,80	61,87	61,95	62,02	62,09	62,17	62,24
84	62,32	62,39	62,46	62,54	62,61	62,69	62,76	62,84	62,91	62,98
85	63,06	63,13	63,21	63,28	63,35	63,43	63,50	63,58	63,65	63,73
86	63,80	63,87	63,95	64,02	64,10	64,17	64,25	64,32	64,39	64,47
87	64,54	64,62	64,69	64,76	64,84	64,91	64,99	65,06	65,14	65,21
88	65,28	65,36	65,43	65,51	65,58	65,65	65,73	65,80	65,88	65,95
89	66,03	66,10	66,17	66,25	66,32	66,40	66,47	66,54	66,62	66,69
90	66,77	66,84	66,92	66,99	67,06	67,14	67,21	67,29	67,36	67,44
91	67,51	67,58	67,66	67,73	67,81	67,88	67,95	68,03	68,10	68,18
92	68,25	68,33	68,40	68,47	68,55	68,62	68,70	68,77	68,84	68,92
93	68,99	69,07	69,14	69,22	69,29	69,36	69,44	69,51	69,59	69,66
94	69,73	69,81	69,88	69,96	70,03	70,11	70,18	70,25	70,33	70,40
95	70,48	70,55	70,63	70,70	70,77	70,85	70,92	71,00	71,07	71,14
96	71,22	71,29	71,37	71,44	71,52	71,59	71,66	71,74	71,81	71,89
97	71,96	72,03	72,11	72,18	72,26	72,33	72,41	72,48	72,55	72,63
98	72,70	72,78	72,85	72,92	73,00	73,07	73,15	73,22	73,30	73,37
99	73,44	73,52	73,59	73,67	73,74	73,82	73,89	73,96	74,04	74,11
100	74,19	74,26	74,33	74,41	74,48	74,56	74,63	74,71	74,78	74,85

mg of Na₂O	1000	2000	3000	4000	5000	6000	7000	8000	9000	10 000
mg of Na⁺	741,9	1483,7	2225,6	2967,4	3709,3	4451,2	5193,0	5934,9	6676,7	7418,6

Table 40

Conversion of Milligrams of CaO into Milligrams of Ca^{2+}

Whole mg of CaO	Tenths of a mg of CaO									
	0	1	2	3	4	5	6	7	8	9
0	—	0,1	0,1	0,2	0,3	0,4	0,4	0,5	0,6	0,6
1	0,71	0,79	0,86	0,93	1,00	1,07	1,14	1,21	1,29	1,36
2	1,43	1,50	1,57	1,64	1,71	1,79	1,86	1,93	2,00	2,07
3	2,14	2,22	2,29	2,36	2,43	2,50	2,57	2,64	2,72	2,79
4	2,86	2,93	3,00	3,07	3,14	3,22	3,29	3,36	3,43	3,50
5	3,57	3,64	3,72	3,79	3,86	3,93	4,00	4,07	4,15	4,22
6	4,29	4,36	4,43	4,50	4,57	4,65	4,72	4,79	4,86	4,93
7	5,00	5,07	5,15	5,22	5,29	5,36	5,43	5,50	5,57	5,65
8	5,72	5,79	5,86	5,93	6,00	6,07	6,15	6,22	6,29	6,36
9	6,43	6,50	6,57	6,65	6,72	6,79	6,86	6,93	7,00	7,08
10	7,15	7,22	7,29	7,36	7,43	7,50	7,58	7,65	7,72	7,79
11	7,86	7,93	8,00	8,08	8,15	8,22	8,29	8,36	8,43	8,50
12	8,58	8,65	8,72	8,79	8,86	8,93	9,00	9,08	9,15	9,22
13	9,29	9,36	9,43	9,50	9,58	9,65	9,72	9,79	9,86	9,93
14	10,01	10,08	10,15	10,22	10,29	10,36	10,43	10,51	10,58	10,65
15	10,72	10,79	10,86	10,94	11,01	11,08	11,15	11,22	11,29	11,36
16	11,44	11,51	11,58	11,65	11,72	11,79	11,86	11,94	12,01	12,08
17	12,15	12,22	12,29	12,36	12,44	12,51	12,58	12,65	12,72	12,79
18	12,86	12,94	13,01	13,08	13,15	13,22	13,29	13,37	13,44	13,51
19	13,58	13,65	13,72	13,79	13,87	13,94	14,01	14,08	14,15	14,22
20	14,29	14,37	14,44	14,51	14,58	14,65	14,72	14,79	14,87	14,94
21	15,01	15,08	15,15	15,22	15,30	15,37	15,44	15,51	15,58	15,65
22	15,72	15,80	15,87	15,94	16,01	16,08	16,15	16,22	16,30	16,37
23	16,44	16,51	16,58	16,65	16,72	16,80	16,87	16,94	17,01	17,08
24	17,15	17,22	17,30	17,37	17,44	17,51	17,58	17,65	17,72	17,79
25	17,87	17,94	18,01	18,09	18,16	18,23	18,30	18,37	18,44	18,51
26	18,58	18,66	18,73	18,80	18,87	18,94	19,01	19,09	19,16	19,23
27	19,30	19,37	19,44	19,51	19,59	19,66	19,73	19,80	19,87	19,94
28	20,01	20,09	20,16	20,23	20,30	20,37	20,44	20,52	20,59	20,66
29	20,73	20,80	20,87	20,94	21,02	21,09	21,16	21,23	21,30	21,37
30	21,44	21,52	21,59	21,66	21,73	21,80	21,87	21,94	22,02	22,09
31	22,16	22,23	22,30	22,37	22,44	22,52	22,59	22,66	22,73	22,80
32	22,87	22,95	23,02	23,09	23,16	23,23	23,30	23,37	23,45	23,52
33	23,59	23,66	23,73	23,80	23,87	23,95	24,02	24,09	24,16	24,23
34	24,30	24,37	24,44	24,51	24,59	24,66	24,73	24,80	24,87	24,94
35	25,01	25,09	25,16	25,23	25,30	25,37	25,44	25,51	25,59	25,66
36	25,73	25,80	25,87	25,94	26,02	26,09	26,16	26,23	26,30	26,37
37	26,44	26,52	26,59	26,66	26,73	26,80	26,87	26,94	27,02	27,09
38	27,16	27,23	27,30	27,37	27,44	27,52	27,59	27,66	27,73	27,80
39	27,87	27,94	28,02	28,09	28,16	28,23	28,30	28,37	28,45	28,52
40	28,59	28,66	28,73	28,80	28,87	28,95	29,02	29,09	29,16	29,23
41	29,30	29,37	29,45	29,52	29,59	29,66	29,73	29,80	29,87	29,95
42	30,02	30,09	30,16	30,23	30,30	30,37	30,45	30,52	30,59	30,66
43	30,73	30,80	30,88	30,95	31,02	31,09	31,16	31,23	31,30	31,38
44	31,45	31,52	31,59	31,66	31,73	31,80	31,88	31,95	32,02	32,09
45	32,16	32,23	32,30	32,38	32,45	32,52	32,59	32,66	32,73	32,80
46	32,88	32,95	33,02	33,09	33,16	33,23	33,31	33,38	33,45	33,52
47	33,59	33,66	33,73	33,81	33,88	33,95	34,02	34,09	34,16	34,23
48	34,31	34,38	34,45	34,52	34,59	34,66	34,73	34,81	34,88	34,95
49	35,02	35,09	35,16	35,23	35,31	35,38	35,45	35,52	35,59	35,66
50	35,73	35,81	35,88	35,95	36,02	36,09	36,16	36,24	36,31	36,38

Whole mg of CaO	Tenths of a mg of CaO									
	0	1	2	3	4	5	6	7	8	9
51	36,45	36,52	36,59	36,66	36,74	36,81	36,88	36,95	37,02	37,09
52	37,16	37,24	37,31	37,38	37,45	37,52	37,59	37,66	37,74	37,81
53	37,88	37,95	38,02	38,09	38,16	38,24	38,31	38,38	38,45	38,52
54	38,59	38,67	38,74	38,81	38,88	38,95	39,02	39,09	39,17	39,24
55	39,31	39,38	39,45	39,52	39,59	39,67	39,74	39,81	39,88	39,95
56	40,02	40,09	40,17	40,24	40,31	40,38	40,45	40,52	40,59	40,67
57	40,74	40,81	40,88	40,95	41,02	41,10	41,17	41,24	41,31	41,38
58	41,45	41,52	41,60	41,67	41,74	41,81	41,88	41,95	42,02	42,10
59	42,17	42,24	42,31	42,38	42,45	42,52	42,60	42,67	42,74	42,81
60	42,88	42,95	43,02	43,10	43,17	43,24	43,31	43,38	43,45	43,53
61	43,60	43,67	43,74	43,81	43,88	43,95	44,03	44,10	44,17	44,24
62	44,31	44,38	44,45	44,53	44,60	44,67	44,74	44,81	44,88	44,95
63	45,03	45,10	45,17	45,24	45,31	45,38	45,45	45,53	45,60	45,67
64	45,74	45,81	45,88	45,96	46,03	46,10	46,17	46,24	46,31	46,38
65	46,46	46,53	46,60	46,67	46,74	46,81	46,88	46,96	47,03	47,10
66	47,17	47,24	47,31	47,38	47,46	47,53	47,60	47,67	47,74	47,81
67	47,88	47,96	48,03	48,10	48,17	48,24	48,31	48,39	48,46	48,53
68	48,60	48,67	48,74	48,81	48,89	48,96	49,03	49,10	49,17	49,24
69	49,31	49,39	49,46	49,53	49,60	49,67	49,74	49,81	49,89	49,96
70	50,03	50,10	50,17	50,24	50,31	50,39	50,46	50,53	50,60	50,67
71	50,74	50,82	50,89	50,96	51,03	51,10	51,17	51,24	51,32	51,39
72	51,46	51,53	51,60	51,67	51,74	51,82	51,89	51,96	52,03	52,10
73	52,17	52,24	52,32	52,39	52,46	52,53	52,60	52,67	52,74	52,82
74	52,89	52,96	53,03	53,10	53,17	53,25	53,32	53,39	53,46	53,53
75	53,60	53,67	53,75	53,82	53,89	53,96	54,03	54,10	54,17	54,25
76	54,32	54,39	54,46	54,53	54,60	54,67	54,75	54,82	54,89	54,96
77	55,03	55,10	55,17	55,25	55,32	55,39	55,46	55,53	55,60	55,68
78	55,75	55,82	55,89	55,96	56,03	56,10	56,18	56,25	56,32	56,39
79	56,46	56,53	56,60	56,68	56,75	56,82	56,89	56,96	57,03	57,10
80	57,18	57,25	57,32	57,39	57,46	57,53	57,60	57,68	57,75	57,82
81	57,89	57,96	58,03	58,11	58,18	58,25	58,32	58,39	58,46	58,53
82	58,61	58,68	58,75	58,82	58,89	58,96	59,03	59,11	59,18	59,25
83	59,32	59,39	59,46	59,53	59,61	59,68	59,75	59,82	59,89	59,96
84	60,03	60,11	60,18	60,25	60,32	60,39	60,46	60,54	60,61	60,68
85	60,75	60,82	60,89	60,96	61,04	61,11	61,18	61,25	61,32	61,39
86	61,46	61,54	61,61	61,68	61,75	61,82	61,89	61,96	62,04	62,11
87	62,18	62,25	62,32	62,39	62,46	62,54	62,61	62,68	62,75	62,82
88	62,89	62,97	63,04	63,11	63,18	63,25	63,32	63,39	63,47	63,54
89	63,61	63,68	63,75	63,82	63,89	63,97	64,04	64,11	64,18	64,25
90	64,32	64,39	64,47	64,54	64,61	64,68	64,75	64,82	64,89	64,97
91	65,04	65,11	65,18	65,25	65 32	65,40	65,47	65,54	65,61	65,68
92	65,75	65,82	65,90	65,97	66,04	66,11	66,18	66,25	66,32	66,40
93	66,47	66,54	66,61	66,68	66,75	66,82	66,90	66,97	67,04	67,11
94	67,18	67,25	67,32	67,40	67,47	67,54	67,61	67,68	67,75	67,83
95	67,90	67,97	68,04	68,11	68,18	68,25	68,33	68,40	68,47	68,54
96	68,61	68,68	68,75	68,83	68,90	68,97	69,04	69,11	69,18	69,25
97	69,33	69,40	69,47	69,54	69,61	69,68	69,75	69,83	69,90	69,97
98	70,04	70,11	70,18	70,25	70,33	70,40	70,47	70,54	70,61	70,68
99	70,76	70,83	70,90	70,97	71,04	71,11	71,18	71,26	71,33	71,40
100	71,47	71,54	71,61	71,68	71,76	71,83	71,90	71,97	72,04	72,11

mg of CaO	1000	2000	3000	4000	5000	6000	7000	8000	9000	10 000
mg of Ca^{2+}	714,7	1429,4	2144,1	2858,8	3573,5	4288,2	5002,9	5717,5	6432,2	7146,9

Table 41

Conversion of Milligrams of MgO into Milligrams of Mg^{2+}

Whole mg of MgO	Tenths of a mg of MgO									
	0	1	2	3	4	5	6	7	8	9
0	—	0,1	0,1	0,2	0,2	0,3	0,4	0,4	0,5	0,5
1	0,60	0,66	0,72	0,78	0,84	0,90	0,97	1,03	1,09	1,15
2	1,21	1,27	1,33	1,39	1,45	1,51	1,57	1,63	1,69	1,75
3	1,81	1,87	1,93	1,99	2,05	2,11	2,17	2,23	2,29	2,35
4	2,41	2,47	2,53	2,59	2,65	2,71	2,77	2,84	2,90	2,96
5	3,02	3,08	3,14	3,20	3,26	3,32	3,38	3,44	3,50	3,56
6	3,62	3,68	3,74	3,80	3,86	3,92	3,98	4,04	4,10	4,16
7	4,22	4,28	4,34	4,40	4,46	4,52	4,58	4,64	4,71	4,77
8	4,83	4,89	4,95	5,01	5,07	5,13	5,19	5,25	5,31	5,37
9	5,43	5,49	5,55	5,61	5,67	5,73	5,79	5,85	5,91	5,97
10	6,03	6,09	6,15	6,21	6,27	6,33	6,39	6,45	6,51	6,57
11	6,64	6,70	6,76	6,82	6,88	6,94	7,00	7,06	7,12	7,18
12	7,24	7,30	7,36	7,42	7,48	7,54	7,60	7,66	7,72	7,78
13	7,84	7,90	7,96	8,02	8,08	8,14	8,20	8,26	8,32	8,38
14	8,44	8,51	8,57	8,63	8,69	8,75	8,81	8,87	8,93	8,99
15	9,05	9,11	9,17	9,23	9,29	9,35	9,41	9,47	9,53	9,59
16	9,65	9,71	9,77	9,83	9,89	9,95	10,01	10,07	10,13	10,19
17	10,25	10,31	10,38	10,44	10,50	10,56	10,62	10,68	10,74	10,80
18	10,86	10,92	10,98	11,04	11,10	11,16	11,22	11,28	11,34	11,40
19	11,46	11,52	11,58	11,64	11,70	11,76	11,82	11,88	11,94	12,00
20	12,06	12,12	12,18	12,25	12,31	12,37	12,43	12,49	12,55	12,61
21	12,67	12,73	12,79	12,85	12,91	12,97	13,03	13,09	13,15	13,21
22	13,27	13,33	13,39	13,45	13,51	13,57	13,63	13,69	13,75	13,81
23	13,87	13,93	13,99	14,05	14,12	14,18	14,24	14,30	14,36	14,42
24	14,48	14,54	14,60	14,66	14,72	14,78	14,84	14,90	14,96	15,02
25	15,08	15,14	15,20	15,26	15,32	15,38	15,44	15,50	15,56	15,62
26	15,68	15,74	15,80	15,86	15,92	15,99	16,05	16,11	16,17	16,23
27	16,29	16,35	16,41	16,47	16,53	16,59	16,65	16,71	16,77	16,83
28	16,89	16,95	17,01	17,07	17,13	17,19	17,25	17,31	17,37	17,43
29	17,49	17,55	17,61	17,67	17,73	17,79	17,85	17,92	17,98	18,04
30	18,10	18,16	18,22	18,28	18,34	18,40	18,46	18,52	18,58	18,64
31	18,70	18,76	18,82	18,88	18,94	19,00	19,06	19,12	19,18	19,24
32	19,30	19,36	19,42	19,48	19,54	19,60	19,66	19,72	19,79	19,85
33	19,91	19,97	20,03	20,09	20,15	20,21	20,27	20,33	20,39	20,45
34	20,51	20,57	20,63	20,69	20,75	20,81	20,87	20,93	20,99	21,05
35	21,11	21,17	21,23	21,29	21,35	21,41	21,47	21,53	21,59	21,65
36	21,72	21,78	21,84	21,90	21,96	22,02	22,08	22,14	22,20	22,26
37	22,32	22,38	22,44	22,50	22,56	22,62	22,68	22,74	22,80	22,86
38	22,92	22,98	23,04	23,10	23,16	23,22	23,28	23,34	23,40	23,46
39	23,52	23,59	23,65	23,71	23,77	23,83	23,89	23,95	24,01	24,07
40	24,13	24,19	24,25	24,31	24,37	24,43	24,49	24,55	24,61	24,67
41	24,73	24,79	24,85	24,91	24,97	25,03	25,09	25,15	25,21	25,27
42	25,33	25,39	25,46	25,52	25,58	25,64	25,70	25,76	25,82	25,88
43	25,94	26,00	26,06	26,12	26,18	26,24	26,30	26,36	26,42	26,48
44	26,54	26,60	26,66	26,72	26,78	26,84	26,90	26,96	27,02	27,08
45	27,14	27,20	27,26	27,32	27,39	27,45	27,51	27,57	27,63	27,69
46	27,75	27,81	27,87	27,93	27,99	28,05	28,11	28,17	28,23	28,29
47	28,35	28,41	28,47	28,53	28,59	28,65	28,71	28,77	28,83	29,89
48	28,95	29,01	29,07	29,13	29,19	29,26	29,32	29,38	29,44	29,50
49	29,56	29,62	29,68	29,74	29,80	29,86	29,92	29,98	30,04	30,10
50	30,16	30,22	30,28	30,34	30,40	30,46	30,52	30,58	30,64	30,70

Whole mg of MgO	Tenths of a mg of MgO									
	0	1	2	3	4	5	6	7	8	9
51	30,76	30,82	30,88	30,94	31,00	31,06	31,13	31,19	31,25	31,31
52	31,37	31,43	31,49	31,55	31,61	31,67	31,73	31,79	31,85	31,91
53	31,97	32,03	32,09	32,15	32,21	32,27	32,33	32,39	32,45	32,51
54	32,57	32,63	32,69	32,75	32,81	32,87	32,93	33,00	33,06	33,12
55	33,18	33,24	33,30	33,36	33,42	33,48	33,54	33,60	33,66	33,72
56	33 78	33,84	33,90	33,96	34,02	34,08	34,14	34,20	34,26	34,32
57	34,38	34,44	34,50	34,56	34,62	34,68	34,74	34,80	34.87	34,93
58	34,99	35,05	35,11	35,17	35,23	35,29	35,35	35,41	35,47	35,53
59	35,59	35,65	35,71	35,77	35,83	35,89	35,95	36,01	36,07	36,13
60	36,19	36,25	36,31	36,37	36,43	36,49	36,55	36,61	36,67	36,74
61	36,80	36,86	36,92	36,98	37,04	37,10	37,16	37,22	37,28	37,34
62	37,40	37,46	37,52	37,58	37,64	37,70	37,76	37,82	37,88	37,94
63	38,00	38,06	38,12	38,18	38,24	38,30	38,36	38,42	38,48	38,54
64	38,60	38,67	38,73	38,79	38,85	38,91	38,97	39,03	39,09	39,15
65	39,21	39,27	39,33	39,39	39,45	39,51	39,57	39,63	39,69	39,75
66	39,81	39,87	39,93	39,99	40,05	40,11	40,17	40,23	40,29	40,35
67	40,41	40,47	40,54	40,60	40,66	40,72	40,78	40,84	40,90	40,96
68	41,02	41,06	41,14	41,20	41,26	41,32	41,38	41,44	41,50	41,56
69	41,62	41,68	41,74	41,80	41,86	41,92	41,98	42,04	42,10	42,16
70	42,22	42,28	42,34	42,40	42,47	42,53	42,59	42,65	42,71	42,77
71	42,83	42,89	42,95	43,01	43,07	43,13	43·19	43,25	43,31	43,37
72	43,43	43,49	43,55	43,61	43,67	43,73	43,79	43,85	43,91	43,97
73	44,03	44,09	44,15	44,21	44,27	44,34	44,40	44,46	44,52	44,58
74	44,64	44,70	44,76	44,82	44,88	44,94	45,00	45,06	45,12	45,18
75	45,24	45,30	45,36	45,42	45,48	45,54	45,60	45 66	45,72	45,78
76	45,84	45,90	45,96	46,02	46,08	46,14	46,21	46,27	46,33	46,39
77	46,45	46,51	46,57	46,63	46,69	46,75	46,81	46,87	46,93	46,99
78	47,05	47,11	47,17	47,23	47,29	47.35	47,41	47,47	47,53	47,59
79	47,65	47,71	47,77	47,83	47.89	47,95	48,01	48,08	48,14	48,20
80	48,26	48,32	48,38	48,44	48,50	48,56	48,62	48,68	48,74	48,80
81	48,86	48,92	48,98	49,04	49,10	49,16	49,22	49,28	49.34	49,40
82	49,46	49,52	49,58	49,64	49,70	49,76	49,82	49,88	49,95	50,01
83	50,07	50,13	50,19	50,25	50,31	50,37	50,43	50,49	50,55	50,61
84	50,67	50,73	50,79	50,85	50,91	50,97	51,03	51,09	51,15	51,21
85	51,27	51,33	51,39	51,45	51,51	51,57	51,63	51,69	51,75	51,82
86	51,88	51,94	52,00	52,06	52,12	52,18	52,24	52,30	52,36	52,42
87	52,48	52,54	52,60	52,66	52,72	52,78	52,84	52,90	52,96	53,02
88	53,08	53,14	53,20	53,26	53,32	53,38	53,44	53,50	53,56	53,62
89	53,68	53,75	53,81	53,87	53,93	53,99	54,05	54,11	54,17	54,23
90	54,29	54,35	54,41	54,47	54,53	54,59	54,65	54,71	54,77	54,83
91	54,89	54,95	55,01	55,07	55,13	55,19	55,25	55,31	55,37	55,43
92	55,49	55,55	55,62	55,68	55,74	55,80	55,86	55,92	55,98	56,04
93	56,10	56,16	56,22	56,28	56,34	56,40	56,46	56,52	56,58	56,64
94	56,70	56,76	56,82	56,88	56,94	57,00	57,06	57,12	57,18	57,24
95	57,30	57,36	57,42	57,48	57,55	57,61	57,67	57,73	57,79	57,85
96	57,91	57,97	58,03	58,09	58,15	58,21	58,27	58,33	58,39	58,45
97	58,51	58,57	58,63	58,69	58,75	58,81	58,87	58,93	58,99	59,05
98	59,11	59,17	59,23	59,29	59,35	59,42	59,48	59,54	59,60	59,66
99	59,72	59,78	59,84	59,90	59,96	60,02	60,08	60,14	60,20	60,26
100	60,32	60,38	60,44	60,50	60,56	60,62	60,68	60,74	60,80	60,86

mg of MgO	1000	2000	3000	4000	5000	6000	7000	8000	9000	10 000
mg of Mg^{2+}	603,2	1206,3	1809,5	2412,7	3015,9	3619,0	4222,2	4825,4	5428,6	6031,7

55

Table 42

Conversion of Milligrams of K₂O into Milligrams of K⁺

Whole mg of K_2O	Tenths of a mg of K_2O									
	0	1	2	3	4	5	6	7	8	9
0	—	0,1	0,2	0,2	0,3	0,4	0,5	0,6	0,7	0,7
1	0,83	0,91	1,00	1,08	1,16	1,25	1,33	1,41	1,49	1,58
2	1,66	1,74	1,83	1,91	1,99	2,08	2,16	2,24	2,32	2,41
3	2,49	2,57	2,66	2.74	2,82	2,91	2,99	3,07	3,16	3,24
4	3,32	3,40	3,49	3,57	3,65	3,74	3,82	3,90	3,99	4,07
5	4,15	4,23	4,32	4,40	4,48	4,57	4,65	4,73	4,81	4,90
6	4,98	5,06	5,15	5,23	5,31	5,40	5,48	5,56	5,64	5,73
7	5,81	5,89	5,98	6,06	6,14	6,23	6,31	6,39	6,47	6,56
8	6,64	6,72	6,81	6,89	6,97	7,06	7,14	7,22	7,30	7,39
9	7,47	7,55	7,64	7,72	7,80	7,89	7,97	8,05	8,13	8,22
10	8,30	8,38	8,47	8,55	8,63	8,72	8,80	8,88	8,97	9,05
11	9,13	9,21	9,30	9,38	9,46	9,55	9,63	9,71	9,80	9,88
12	9,96	10,04	10,13	10,21	10,29	10,38	10,46	10,54	10,63	10,71
13	10,79	10,87	10,96	11,04	11,12	11,21	11,29	11,37	11,46	11,54
14	11,62	11,70	11,79	11,87	11,95	12,04	12,12	12,20	12,29	12,37
15	12,45	12,53	12,62	12,70	12,78	12,87	12,95	13,03	13,12	13,20
16	13,28	13,36	13,45	13,53	13,61	13,70	13,79	13,86	13,95	14,03
17	14,11	14,19	14,28	14,36	14,44	14,53	14,61	14,69	14,78	14,86
18	14,94	15,02	15,11	15,19	15,27	15,36	15,44	15,52	15,61	15,69
19	15,77	15,85	15,94	16,02	16,10	16,19	16,27	16,35	16,44	16,52
20	16,60	16,69	16,77	16,85	16,93	17,02	17,10	17,18	17,27	17,35
21	17,43	17,52	17,60	17,68	17,76	17,85	17,93	18,01	18,10	18,18
22	18,26	18,35	18,43	18,51	18,59	18,68	18,76	18,84	18,93	19,01
23	19,09	19,18	19,26	19,34	19,42	19,51	19,59	19,67	19,76	19,84
24	19,92	20,01	20,09	20,17	20,25	20,34	20,42	20,50	20,59	20,67
25	20,75	20,84	20,92	21,00	21,08	21,17	21,25	21,33	21,42	21,50
26	21,58	21,67	21,75	21,83	21,91	22,00	22,08	22,16	22,25	22,33
27	22,41	22,50	22,58	22,66	22,74	22,83	22,91	22,99	23,08	23,16
28	23,24	23,33	23,41	23,49	23,57	23,66	23,74	23,82	23,91	23,99
29	24,07	24,16	24,24	24,32	24,40	24,49	24,57	24,65	24,74	24,82
30	24,90	24,99	25,07	25,15	25,24	25,32	25,40	25,48	25,57	25,65
31	25,73	25,82	25,90	25,98	26,07	26,15	26,23	26,31	26,40	26,48
32	26,56	26,65	26,73	26,81	26,90	26,98	27,06	27,14	27,23	27,31
33	27,39	27,48	27,56	27,64	27,73	27,81	27,89	27,97	28,06	28,14
34	28,22	28,31	28,39	28,47	28,56	28,64	28,72	28,80	28,89	28,97
35	29,05	29,14	29,22	29,30	29,39	29,47	29,55	29,63	29,72	29,80
36	29,88	29,97	30,05	30,13	30,22	30,30	30,38	30,46	30,55	30,63
37	30,71	30,80	30,88	30,96	31,05	31,13	31,21	31,29	31,38	31,46
38	31,55	31,64	31,72	31,80	31,89	31,97	32,05	32,13	32,22	32,30
39	32,38	32,47	32,55	32,63	32,72	32,80	32,88	32,96	33,05	33,13
40	33,21	33,29	33,37	33,45	33,54	33,62	33,70	33,79	33,87	33,95
41	34,04	34,12	34,20	34,28	34,37	34,45	34,53	34,62	34,70	34,78
42	34,87	34,95	35,03	35,11	35,20	35,28	35,36	35,45	35,53	35,61
43	35,70	35,78	35,86	35,94	36,03	36,11	36,19	36,28	36,36	36,44
44	36,53	36,61	36,69	36,77	36,86	36,94	37,02	37,11	37,19	37,27
45	37,36	37,44	37,52	37,60	37,69	37,77	37,85	37,94	38,02	38,10
46	38,19	38,27	38,35	38,43	38,52	38,60	38,68	38,77	38,85	38,93
47	39,02	39,10	39,18	39,27	39,35	39,43	39,51	39,60	39,68	39,76
48	39,85	39,93	40,01	40,10	40,18	40,26	40,34	40,43	40,51	40,59
49	40,68	40,76	40,84	40,93	41,01	41,09	41,17	41,26	41,34	41,42
50	41,51	41,59	41,67	41,76	41,84	41,92	42,00	42,09	42,17	42,25

Table 43

Conversion of Milligrams of FeO into Milligrams of Fe^{2+}

Whole mg of FeO	Tenths of a mg of FeO									
	0	1	2	3	4	5	6	7	8	9
0	—	0,1	0,2	0,2	0,3	0,4	0,5	0,5	0,6	0,7
1	0,78	0,85	0,93	1,01	1,09	1,17	1,24	1,32	1,40	1,48
2	1,55	1,63	1,71	1,79	1,87	1,94	2,02	2,10	2,18	2,25
3	2,33	2,41	2,49	2,56	2,64	2,72	2,80	2,88	2,95	3,03
4	3,11	3,19	3,26	3,34	3,42	3,50	3,58	3,65	3,73	3,81
5	3,89	3,96	4,04	4,12	4,20	4,28	4,35	4,43	4,51	4,59
6	4,66	4,74	4,82	4,90	4,97	5,05	5,13	5,21	5,29	5,36
7	5,44	5,52	5,60	5,67	5,75	5,83	5,91	5,99	6,06	6,14
8	6,22	6,30	6,37	6,45	6,53	6,61	6,68	6,76	6,84	6,92
9	7,00	7,07	7,15	7,23	7,31	7,38	7,46	7,54	7,62	7,70
10	7,77	7,85	7,93	8,01	8,08	8,16	8,24	8,32	8,39	8,47
11	8,55	8,63	8,71	8,78	8,86	8,94	9,02	9,09	9,17	9,25
12	9,33	9,41	9,48	9,56	9,64	9,72	9,79	9,87	9,95	10,03
13	10,10	10,18	10,26	10,34	10,42	10,49	10,57	10,65	10,73	10,80
14	10,88	10,96	11,04	11,12	11,19	11,27	11,35	11,43	11,50	11,58
15	11,66	11,74	11,81	11,89	11,97	12,05	12,13	12,20	12,28	12,36
16	12,44	12,51	12,59	12,67	12,75	12,83	12,90	12,98	13,06	13,14
17	13,21	13,29	13,37	13,45	13,53	13,60	13,68	13,76	13,84	13,91
18	13,99	14,07	14,15	14,22	14,30	14,38	14,46	14,54	14,61	14,69
19	14,77	14,85	14,92	15,00	15,08	15,16	15,24	15,31	15,39	15,47
20	15,55	15,62	15,70	15,78	15,86	15,93	16,01	16,09	16,17	16,25
21	16,32	16,40	16,48	16,56	16,63	16,71	16,79	16,87	16,95	17,02
22	17,10	17,18	17,26	17,33	17,41	17,49	17,57	17,64	17,72	17,80
23	17,88	17,96	18,03	18,11	18,19	18,27	18,34	18,42	18,50	18,58
24	18,66	18,73	18,81	18,89	18,97	19,04	19,12	19,20	19,28	19,35
25	19,43	19,51	19,59	19,67	19,74	19,82	19,90	19,98	20,05	20,13
26	20,21	20,29	20,36	20,44	20,52	20,60	20,68	20,75	20,83	20,91
27	20,99	21,06	21,14	21,22	21,30	21,38	21,45	21,53	21,61	21,69
28	21,76	21,84	21,92	22,00	22,08	22,15	22,23	22,31	22,39	22,46
29	22,54	22,62	22,70	22,77	22,85	22,93	23,01	23,09	23,16	23,24
30	23,32	23,40	23,47	23,55	23,63	23,71	23,79	23,86	23,94	24,02
31	24,10	24,17	24,25	24,33	24,41	24,48	24,56	24,64	24,72	24,80
32	24,87	24,95	25,03	25,11	25,18	25,26	25,34	25,42	25,50	25,57
33	25,65	25,73	25,81	25,88	25,96	26,04	26,12	26,20	26,27	26,35
34	26,43	26,51	26,58	26,66	26,74	26,82	26,89	26,97	27,05	27,13
35	27,21	27,28	27,36	27,44	27,52	27,59	27,67	27,75	27,83	27,91
36	27,98	28,06	28,14	28,22	28,29	28,37	28,45	28,53	28,60	28,68
37	28,76	28,84	28,92	28,99	29,07	29,15	29,23	29,30	29,38	29,46
38	29,54	29,62	29,69	29,77	29,85	29,92	30,00	30,08	30,16	30,24
39	30,31	30,39	30,47	30,55	30,63	30,70	30,78	30,86	30,94	31,01
40	31,09	31,17	31,25	31,33	31,40	31,48	31,56	31,64	31,71	31,79
41	31,87	31,95	32,02	32,10	32,18	32,26	32,34	32,41	32,49	32,57
42	32,65	32,72	32,80	32,88	32,96	33,04	33,11	33,19	33,27	33,35
43	33,42	33,50	33,58	33,66	33,73	33,81	33,89	33,97	34,05	34,12
44	34,20	34,28	34,36	34,43	34,51	34,59	34,67	34,75	34,82	34,90
45	34,98	35,06	35,13	35,21	35,29	35,37	35,44	35,52	35,60	35,68
46	35,76	35,83	35,91	35,99	36,07	36,14	36,22	36,30	36,38	36,46
47	36,53	36,61	36,69	36,77	36,84	36,92	37,00	37,08	37,15	37,23
48	37,31	37,38	37,47	37,54	37,62	37,70	37,78	37,85	37,93	38,01
49	38,09	38,17	38,24	38,32	38,40	38,48	38,55	38,63	38,71	38,79
50	38,87	38,94	39,02	39,10	39,18	39,25	39,33	39,41	39,49	39,56

Table 44

Conversion of Milligrams of Fe_2O_3 into Milligrams of Fe^{3+}

Whole mg of Fe_2O_3	Tenths of a mg of Fe_2O_3									
	0	1	2	3	4	5	6	7	8	9
0	—	0,1	0,1	0,21	0,3	0,3	0,4	0,5	0,6	0,6
1	0,70	0,77	0,84	0,91	0,98	1,05	1,12	1,19	1,26	1,33
2	1,40	1,47	1,54	1,61	1,68	1,75	1,82	1,89	1,96	2,03
3	2,10	2,17	2,24	2,31	2,38	2,45	2,52	2,59	2,66	2,73
4	2,80	2,87	2,94	3,01	3,08	3,15	3,22	3,29	3,36	3,43
5	3,50	3,57	3,64	3,71	3,78	3,85	3,92	3,99	4,06	4,13
6	4,20	4,27	4,34	4,41	4,48	4,55	4,62	4,69	4,76	4,83
7	4,90	4,97	5,04	5,11	5,18	5,25	5,32	5,39	5,46	5,53
8	5,60	5,67	5,74	5,81	5,88	5,95	6,02	6,09	6,16	6,22
9	6,29	6,36	6,43	6,50	6,57	6,64	6,71	6,78	6,85	6,92
10	6,99	7,06	7,13	7,20	7,27	7,34	7,41	7,48	7,55	7,62
11	7,69	7,76	7,83	7,90	7,97	8,04	8,11	8,18	8,25	8,32
12	8,39	8,46	8,53	8,60	8,67	8,74	8,81	8,88	8,95	9,02
13	9,09	9,16	9,23	9,30	9,37	9,44	9,51	9,58	9,65	9,72
14	9,79	9,86	9,93	10,00	10,07	10,14	10,21	10,28	10,35	10,42
15	10,49	10,56	10,63	10,70	10,77	10,84	10,91	10,98	11,05	11,12
16	11,19	11,26	11,33	11,40	11,47	11,54	11,61	11,68	11,75	11,82
17	11,89	11,96	12,03	12,10	12,17	12,24	12,31	12,38	12,45	12,52
18	12,59	12,66	12,73	12,80	12,87	12,94	13,01	13,08	13,15	13,22
19	13,29	13,36	13,43	13,50	13,57	13,64	13,71	13,78	13,85	13,92
20	13,99	14,06	14,13	14,20	14,27	14,34	14,41	14,48	14,55	14,62
21	14,69	14,76	14,83	14,90	14,97	15,04	15,11	15,18	15,25	15,32
22	15,39	15,46	15,53	15,60	15,67	15,74	15,81	15,88	15,95	16,02
23	16,09	16,16	16,23	16,30	16,37	16,44	16,51	16,58	16,65	16,72
24	16,79	16,86	16,93	17,00	17,07	17,14	17,21	17,28	17,35	17,42
25	17,49	17,56	17,63	17,70	17,77	17,84	17,91	17,98	18,05	18,12
26	18,19	18,26	18,33	18,40	18,47	18,54	18,61	18,68	18,74	18,81
27	18,88	18,95	19,02	19,09	19,16	19,23	19,30	19,37	19,44	19,51
28	19,58	19,65	19,72	19,79	19,86	19,93	20,00	20,07	20,14	20,21
29	20,28	20,35	20,42	20,49	20,56	20,63	20,70	20,77	20,84	20,91
30	20,98	21,05	21,12	21,19	21,26	21,33	21,40	21,47	21,54	21,61
31	21,68	21,75	21,82	21,89	21,96	22,03	22,10	22,17	22,24	22,31
32	22,38	22,45	22,52	22,59	22,66	22,73	22,80	22,87	22,94	23,01
33	23,08	23,15	23,22	23,29	23,36	23,43	23,50	23,57	23,64	23,71
34	23,78	23,85	23,92	23,99	24,06	24,13	24,20	24,27	24,34	24,41
35	24,48	24,55	24,62	24,69	24,76	24,83	24,90	24,97	25,04	25,11
36	25,18	25,25	25,32	25,39	25,46	25,53	25,60	25,67	25,74	25,81
37	25,88	25,95	26,02	26,09	26,16	26,23	26,30	26,37	26,44	26,51
38	26,58	26,65	26,72	26,79	26,86	26,93	27,00	27,07	27,14	27,21
39	27,28	27,35	27,42	27,49	27,56	27,63	27,70	27,77	27,84	27,91
40	27,98	28,05	28,12	28,19	28,26	28,33	28,40	28,47	28,54	28,61
41	28,68	28,75	28,82	28,89	28,96	29,03	29,10	29,17	29,24	29,31
42	29,38	29,45	29,52	29,59	29,66	29,73	29,80	29,87	29,94	30,01
43	30,08	30,15	30,22	30,29	30,36	30,43	30,50	30,57	30,64	30,71
44	30,78	30,85	30,92	30,99	31,06	31,13	31,20	31,26	31,33	31,40
45	31,47	31,54	31,61	31,68	31,75	31,82	31,89	31,96	32,03	32,10
46	32,17	32,24	32,31	32,38	32,45	32,52	32,59	32,66	32,73	32,80
47	32,87	32,94	33,01	33,08	33,15	33,22	33,29	33,36	33,43	33,50
48	33,57	33,64	33,71	33,78	33,85	33,92	33,99	34,06	34,13	34,20
49	34,27	34,34	34,41	34,48	34,55	34,62	34,69	34,76	34,83	34,90
50	34,97	35,04	35,11	35,18	35,25	35,32	35,39	35,46	35,53	35,60

Table 45

Conversion of Milligrams of Al_2O_3 into Milligrams of Al^{3+}

Whole mg of Al_2O_3	Tenths of a mg of Al_2O_3									
	0	1	2	3	4	5	6	7	8	9
0	—	0,1	0,1	0,2	0,2	0,3	0,3	0,4	0,4	0,5
1	0,53	0,58	0,64	0,69	0,74	0,79	0,85	0,90	0,95	1,01
2	1,06	1,11	1,16	1,22	1,27	1,32	1,38	1,43	1,48	1,53
3	1,59	1,64	1,69	1,75	1,80	1,85	1,91	1,96	2,01	2,06
4	2,12	2,17	2,22	2,28	2,33	2,38	2,43	2,49	2,54	2,59
5	2,65	2,70	2,75	2,80	2,86	2,91	2,96	3,02	3,07	3,12
6	3.18	3,23	3,28	3,33	3,39	3,44	3,49	3,55	3,60	3,65
7	3,70	3,76	3,81	3,86	3,92	3,97	4,02	4,08	4,13	4,18
8	4,23	4,29	4,34	4,39	4,45	4,50	4,55	4,60	4,66	4,71
9	4,76	4,82	4,87	4,92	4,97	5,03	5,08	5,13	5,19	5,24
10	5,29	5,35	5,40	5,45	5,50	5,56	5,61	5,66	5,72	5,77
11	5,82	5,87	5,93	5,98	6,03	6,09	6,14	6,19	6,24	6,30
12	6,35	6,40	6,46	6,51	6,56	6,62	6,67	6,72	6,77	6,83
13	6,88	6,93	6,99	7,04	7,09	7,14	7,20	7,25	7,30	7,36
14	7,41	7,46	7,52	7,57	7,62	7,67	7,73	7,78	7,83	7,89
15	7,94	7,99	8,04	8,10	8,15	8,20	8,26	8,31	8,36	8,41
16	8,47	8,52	8,57	8,63	8,68	8,73	8,79	8,84	8,89	8,94
17	9,00	9,05	9,10	9,16	9,21	9,26	9,31	9,37	9,42	9,47
18	9,53	9,58	9,63	9,68	9,74	9,79	9,84	9,90	9,95	10,00
19	10,06	10,11	10,16	10,21	10,27	10,32	10,37	10,43	10,48	10,53
20	10,58	10,64	10,69	10,74	10,80	10,85	10,90	10,95	11,01	11,06
21	11,11	11,17	11,22	11,27	11,33	11,38	11,43	11,48	11,54	11,59
22	11,64	11,70	11,75	11,80	11,85	11,91	11,96	12,01	12,07	12,12
23	12,17	12,23	12,28	12,33	12,38	12,44	12,49	12,54	12,60	12,65
24	12,70	12,75	12,81	12,86	12,91	12,97	13,02	13,07	13,12	13,18
25	13,23	13,28	13,34	13,39	13,44	13,50	13,55	13,60	13,65	13,71
26	13,76	13,81	13,87	13,92	13,97	14,02	14,08	14,13	14,18	14,24
27	14,29	14,34	14,39	14,45	14,50	14,55	14,61	14,66	14,71	14,77
28	14,82	14,87	14,92	14,98	15,03	15,08	15,14	15,19	15,24	15,29
29	15,35	15,40	15,45	15,51	15,56	15,61	15,67	15,72	15,77	15,82
30	15,88	15,93	15,98	16,04	16,09	16,14	16,19	16,25	16,30	16,35
31	16,41	16,46	16,51	16,56	16,62	16,67	16,72	16,78	16,83	16,88
32	16,94	16,99	17,04	17,09	17,15	17,20	17,25	17,31	17,36	17,41
33	17,46	17,52	17,57	17,62	17,68	17,73	17,78	17,83	17,89	17,94
34	17,99	18,05	18,10	18,15	18,21	18,26	18,31	18,36	18,42	18,47
35	18,52	18,58	18.63	18,68	18,73	18,79	18,84	18,89	18,95	19,00
36	19,05	19,11	19,16	19,21	19,26	19,32	19,37	19,42	19,48	19,53
37	19,58	19,63	19,69	19,74	19,79	19,85	19,90	19,95	20,00	20,06
38	20,11	20,16	20,22	20,27	20,32	20,38	20,43	20,48	20,53	20,59
39	20,64	20,69	20,75	20,80	20,85	20,90	20,96	21,01	21,06	21,12
40	21,17	21,22	21,27	21,33	21,38	21,43	21,49	21,54	21,59	21,65
41	21,70	21,75	21,80	21,86	21,91	21,96	22,02	22,07	22,12	22,17
42	22,23	22,28	22,33	22,39	22,44	22,49	22,55	22,60	22,65	22,70
43	22,76	22,81	22,86	22,92	22,97	23,02	23,07	23,13	23,18	23,23
44	23,29	23,34	23,39	23,44	23,50	23,55	23,60	23,66	23,71	23,76
45	23,82	23,87	23,92	23,97	24,03	24,08	24,13	24,19	24,24	24,29
46	24,34	24,40	24,45	24,50	24,56	24,61	24,66	24,71	24,77	24,82
47	24,87	24,93	24,98	25,03	25,09	25,14	25,19	25,24	25,30	25,35
48	25,40	25,46	25,51	25,56	25,61	25,67	25,72	25,77	25,83	25,88
49	25,93	25,99	26,04	26,09	26,14	26,20	26,25	26,30	26,36	26,41
50	26,46	26,51	26,57	26,62	26,67	26,73	26,78	26,83	26,88	26,94

Table 46

Conversion of Milligrams of MnO into Milligrams of Mn^{2+}

Whole mg of MnO	Tenths of a mg of MnO									
	0	1	2	3	4	5	6	7	8	9
0	—	0,1	0,2	0,2	0,3	0,4	0,5	0,5	0,6	0,7
1	0,77	0,85	0,93	1,01	1,08	1,16	1,24	1,32	1,39	1,47
2	1,55	1,63	1,70	1,78	1,86	1,94	2,01	2,09	2,17	2,25
3	2,32	2,40	2,48	2,56	2,63	2,71	2,79	2,87	2,94	3,02
4	3,10	3,18	3,25	3,33	3,41	3,49	3,56	3,64	3,72	3,79
5	3,87	3,95	4,03	4,10	4,18	4,26	4,34	4,41	4,49	4,57
6	4,65	4,72	4,80	4,88	4,96	5,03	5,11	5,19	5,27	5,34
7	5,42	5,50	5,58	5,65	5,73	5,81	5,89	5,96	6,04	6,12
8	6,20	6,27	6,35	6,43	6,51	6,58	6,66	6,74	6,82	6,89
9	6,97	7,05	7,12	7,20	7,28	7,36	7,43	7,51	7,59	7,67
10	7,74	7,82	7,90	7,98	8,05	8,13	8,21	8,29	8,36	8,44
11	8,52	8,60	8,67	8,75	8,83	8,91	8,98	9,06	9,14	9,22
12	9,29	9,37	9,45	9,53	9,60	9,68	9,76	9,84	9,91	9,99
13	10,07	10,15	10,22	10,30	10,38	10,46	10,53	10,61	10,69	10,76
14	10,84	10,92	11,00	11,07	11,15	11,23	11,31	11,38	11,46	11,54
15	11,62	11,69	11,77	11,85	11,93	12,00	12,08	12,16	12,24	12,31
16	12,39	12,47	12,55	12,62	12,70	12,78	12,86	12,93	13,01	13,09
17	13,17	13,24	13,32	13,40	13,48	13,55	13,63	13,71	13,79	13,86
18	13,94	14,02	14,10	14,17	14,25	14,33	14,40	14,48	14,56	14,64
19	14,71	14,79	14,87	14,95	15,02	15,10	15,18	15,26	15,33	15,41
20	15,49	15,57	15,64	15,72	15,80	15,88	15,95	16,03	16,11	16,19
21	16,26	16,34	16,42	16,50	16,57	16,65	16,73	16,81	16,88	16,96
22	17,04	17,12	17,19	17,27	17,35	17,43	17,50	17,58	17,66	17,73
23	17,81	17,89	17,97	18,04	18,12	18,20	18,28	18,35	18,43	18,51
24	18,59	18,66	18,74	18,82	18,90	18,97	19,05	19,13	19,21	19,28
25	19,36	19,44	19,52	19,59	19,67	19,75	19,83	19,90	19,98	20,06
26	20,14	20,21	20,29	20,37	20,45	20,52	20,60	20,68	20,76	20,83
27	20,91	20,99	21,07	21,14	21,22	21,30	21,37	21,45	21,53	21,61
28	21,68	21,76	21,84	21,92	21,99	22,07	22,15	22,23	22,30	22,38
29	22,46	22,54	22,61	22,69	22,77	22,85	22,92	23,00	23,08	23,16
30	23,23	23,31	23,39	23,47	23,54	23,62	23,70	23,78	23,85	23,93
31	24,01	24,09	24,16	24,24	24,32	24,40	24,47	24,55	24,63	24,71
32	24,78	24,86	24,94	25,01	25,09	25,17	25,25	25,32	25,40	25,48
33	25,56	25,63	25,71	25,79	25,87	25,94	26,02	26,10	26,18	26,25
34	26,33	26,41	26,49	26,56	26,64	26,72	26,80	26,87	26,95	27,03
35	27,11	27,18	27,26	27,34	27,42	27,49	27,57	27,65	27,73	27,80
36	27,88	27,96	28,04	28,11	28,19	28,27	28,35	28,42	28,50	28,58
37	28,65	28,73	28,81	28,89	28,96	29,04	29,12	29,20	29,27	29,35
38	29,43	29,51	29,58	29,66	29,74	29,82	29,89	29,97	30,05	30,13
39	30,20	30,28	30,36	30,44	30,51	30,59	30,67	30,75	30,82	30,90
40	30,98	31,06	31,13	31,21	31,29	31,37	31,44	31,52	31,60	31,68
41	31,75	31,83	31,91	31,98	32,06	32,14	32,22	32,29	32,37	32,45
42	32,53	32,60	32,68	32,76	32,84	32,91	32,99	33,07	33,15	33,22
43	33,30	33,38	33,46	33,53	33,61	33,69	33,77	33,84	33,92	34,00
44	34,08	34,15	34,23	34,31	34,39	34,46	34,54	34,62	34,70	34,77
45	34,85	34,93	35,01	35,08	35,16	35,24	35,32	35,39	35,47	35,55
46	35,62	35,70	35,78	35,86	35,93	36,01	36,09	36,17	36,24	36,32
47	36,40	36,48	36,55	36,63	36,71	36,79	36,86	36,94	37,02	37,10
48	37,17	37,25	37,33	37,41	37,48	37,56	37,64	37,72	37,79	37,87
49	37,95	38,03	38,10	38,18	38,26	38,34	38,41	38,49	38,57	38,65
50	38,72	38,80	38,88	38,96	39,03	39,11	39,19	39,26	39,34	39,42

Table 47

Conversion of Milligrams of N_2O_3 into Milligrams of NO_2^-

Whole mg of N_2O_3	Tenths of a mg of N_2O_3									
	0	1	2	3	4	5	6	7	8	9
0	—	0,1	0,2	0,4	0,5	0,6	0,7	0,8	1,0	1,1
1	1,21	1,33	1,45	1,57	1,69	1,82	1,94	2,06	2,18	2,30
2	2,42	2,54	2,66	2,78	2,90	3,03	3,15	3,27	3,39	3,51
3	3,63	3,75	3,87	3,99	4,11	4,24	4,36	4,48	4,60	4,72
4	4,84	4,96	5,08	5,20	5,32	5,45	5,57	5,69	5,81	5,93
5	6,05	6,17	6,29	6,41	6,53	6,66	6,78	6,90	7,02	7,14
6	7,26	7,38	7,50	7,62	7,74	7,87	7,99	8,11	8,23	8,35
7	8,47	8,59	8,71	8,83	8,95	9,08	9,20	9,32	9,44	9,56
8	9,68	9,80	9,92	10,04	10,16	10,29	10,41	10,53	10,65	10,77
9	10,89	11,01	11,13	11,25	11,37	11,50	11,62	11,74	11,86	11,98
10	12,10	12,22	12,34	12,46	12,58	12,71	12,83	12,95	13,07	13,19

Table 48

Conversion of Milligrams of N_2O_5 into Milligrams of NO_3^-

Whole mg of N_2O_5	Tenths of a mg of N_2O_5									
	0	1	2	3	4	5	6	7	8	9
0	—	0,1	0,2	0,3	0,5	0,6	0,7	0,8	0,9	1,0
1	1,15	1,25	1,38	1,49	1,61	1,72	1,84	1,95	2,07	2,18
2	2,30	2,41	2,53	2,64	2,75	2,87	2,98	3,10	3,21	3,33
3	3,44	3,56	3,67	3,79	3,90	4,02	4,13	4,25	4,36	4,48
4	4,59	4,71	4,82	4,94	5,05	5,17	5,28	5,40	5,51	5,63
5	5,74	5,86	5,97	6,09	6,20	6,31	6,43	6,54	6,66	6,77
6	6,89	7,00	7,12	7,23	7,35	7,46	7,58	7,69	7,81	7,92
7	8,04	8,15	8,27	8,38	8,50	8,61	8,73	8,84	8,96	9,07
8	9,18	9,30	9,41	9,53	9,64	9,76	9,87	9,99	10,10	10,12
9	10,33	10,45	10,56	10,68	10,79	10,91	11,02	11,14	11,25	11,37
10	11,48	11,60	11,71	11,83	11,94	12,06	12,17	12,28	12,40	12,51
11	12,63	12,74	12,86	12,97	13,09	13,20	13,32	13,43	13,55	13,66
12	13,78	13,99	14,01	14,12	14,24	14,35	14,47	14,58	14,70	14,81
13	14.93	15,04	15,16	15,27	15,38	15,50	15,61	15,73	15,84	15,96
14	16,07	16,19	16,30	16,42	16,53	16,65	16,76	16,88	16,99	17,11
15	17,22	17,34	17,45	17,57	17,68	17,80	17,91	18,03	18,14	18,26
16	18,37	18,48	18,60	18,71	18,83	18,94	19,06	19,17	19,29	19,40
17	19,52	19,63	19,75	19,86	19,98	20,09	20,21	20,32	20,44	20,55
18	20,67	20,78	20,90	21,01	21,13	21,24	21,36	21,47	21,58	21,70
19	21,81	21,93	22,04	22,16	22,27	22,39	22,50	22,62	22,73	22,85
20	22,96	23,08	23,19	23,31	23,42	23,54	23,65	23,77	23,88	24,00
21	24,11	24,23	24,34	24,45	24,57	24,68	24,80	24,91	25,03	25,14
22	25,26	25,37	25,49	25,60	25,72	25,83	25,95	26,06	26,18	26,29
23	26,41	26,52	26,64	26,75	26,87	26,98	27,10	27,21	27,33	27,44
24	27.55	27,67	27,78	27,90	28,01	28,13	28,24	28,36	28,47	28,59
25	28,70	28,82	28,93	29,05	29,16	29,28	29,39	29,51	29,62	29,74
26	29,85	29,97	30,08	30,20	30,31	30,43	30,54	30,65	30,77	30,88
27	31,00	31,11	31,23	31,34	31,46	31,57	31,69	31,80	31,92	32,03
28	32,15	32,26	32,38	32,49	32,61	32,72	32,84	32,95	33,07	33,18
29	33,30	33,41	33,53	33,64	33,75	33,87	33,98	34,10	34,21	34,33
30	34,44	35,56	34,67	34,79	34,90	35,02	35,13	35,25	35,36	35,48
31	35,59	35,71	35,82	35,94	36,05	36,17	36,28	36,40	36,51	36,63
32	36,74	36,85	36,97	37,08	37,20	37,31	37,43	37,54	37,66	37,77
33	37,89	38,00	38,12	38,23	38,35	38,46	38,58	38,69	38.81	38,92
34	39,04	39,15	39,27	39,38	39,50	39,61	39,72	39,84	39,95	40,07
35	40,18	40,30	40,41	40,53	40,64	40,76	40,87	40,99	41,10	41,22
36	41,33	41,45	41,56	41,68	41,79	41,91	42,02	42,14	42,25	42,37
37	42,49	42,60	42,71	42,82	42,94	43,05	43,17	43,28	43,40	43,51
38	43,63	43,74	43,86	43,97	44,09	44,20	44,32	44,43	44,55	44,66
39	44,78	44,89	45,01	45,12	45,24	45,35	45,47	45,58	45,70	45,81
40	45,92	46,04	46,15	46,27	46,38	46,50	46,61	46,73	46,84	46,96
41	47,07	47,19	47,30	47,42	47,53	47,65	47,76	47,88	47,99	48,11
42	48,22	48,34	48,45	48,57	48,68	48,80	48,91	49,02	49,14	49,25
43	49,37	49,48	49,60	49,71	49,83	49,94	50,06	50,17	50,29	50,40
44	50,52	50,63	50,75	50,86	50,98	51,09	51,21	51,32	51,44	51,55
45	51,67	51,78	51,90	52,01	52,12	52,24	52,35	52,47	52,58	52,70
46	52,81	52,93	53,04	53,16	53,27	53,39	53,50	53,62	53,73	53,85
47	53,96	54,08	54,19	54,31	54,42	54,54	54,65	54,77	54,88	54,99
48	55,11	55,22	55,34	55,45	55,57	55,68	55,80	55,91	56,03	56,14
49	56,26	56,37	56,49	56,60	56,72	56,83	56,95	57,06	57,18	57,29
50	57,41	57,52	57,63	57,75	57,87	57,98	58,09	58,21	58,32	58,44

Whole mg of N_2O_5	\multicolumn{10}{c}{Tenths of a mg of N_2O_5}
	0	1	2	3	4	5	6	7	8	9
51	58,55	58,67	58,78	58,90	59,01	59,13	59,24	59,36	59,47	59,59
52	59,70	59,82	59,93	60,05	60,16	60,28	60,39	60,51	60,62	60,74
53	60,85	60,97	61,08	61,19	61,31	61,42	61,54	61,65	61,77	61,88
54	62,00	62,11	62,23	62,34	62,46	62,57	62,69	62,80	62,92	63,03
55	63,15	63,26	63,38	63,49	63,61	63,72	63,84	63,95	64,07	64,18
56	64,29	64,41	64,52	64,64	64,75	64,87	64,98	65,10	65,21	65,33
57	65,44	65,56	65,67	65,79	65,90	66,02	66,13	66,25	66,36	66,48
58	66,59	66,71	66,82	66,94	67,05	67,17	67,28	67,39	67,51	67,62
59	67,74	67,85	67,97	68,08	68,20	68,31	68,43	68,54	68,66	68,77
60	68,89	69,00	69,12	69,23	69,35	69,46	69,58	69,69	69,81	69,92
61	70,04	70,15	70,26	70,38	70,49	70,61	70,72	70,84	70,95	71,07
62	71,18	71,30	71,41	71,53	71,64	71,76	71,87	71,99	72,10	72,22
63	72,33	72,45	72,56	72,68	72,79	72,91	73,02	73,14	73,25	73,36
64	73,48	73,59	73,71	73,82	73,94	74,05	74,17	74,28	74,40	74,51
65	74,63	74,74	74,86	74,97	75,09	75,20	75,32	75,42	75,55	75,66
66	75,78	75,89	76,01	76,12	76,24	76,35	76,46	76,58	76,69	76,81
67	76,92	77,04	77,15	77,27	77,38	77,50	77,61	77,73	77,84	77,96
68	78,07	78,19	78,30	78,42	78,53	78,65	78,76	78,88	78,99	79,11
69	79,22	79,34	79,45	79,56	79,68	79,79	79,91	80,02	80,14	80,25
70	80,37	80,48	80,60	80,71	80,83	80,94	81,06	81,17	81,29	81,40
71	81,52	81,63	81,75	81,86	81,98	82,09	82,21	82,32	82,44	82,55
72	82,66	82,78	82,89	83,01	83,12	83,24	83,35	83,47	83,58	83,70
73	83,81	83,93	84,04	84,16	84,27	84,39	84,51	84,62	84,73	84,85
74	84,96	85,08	85,19	85,31	85,42	85,53	85,65	85,76	85,88	85,99
75	86,11	86,22	86,34	86,45	86,57	86,68	86,80	86,91	87,02	87,14
76	87,26	87,37	87,49	87,60	87,72	87,83	87,95	88,06	88,18	88,29
77	88,41	88,52	88,63	88,75	88,86	88,98	89,09	89,21	89,32	89,44
78	89,55	89,67	89,78	89,89	90,01	90,13	90,24	90,36	90,47	90,59
79	90,70	90,82	90,93	91,05	91,16	91,28	91,39	91,51	91,62	91,74
80	91,85	91,96	92,08	92,19	92,31	92,42	92,54	92,65	92,77	92,88
81	93,00	93,11	93,23	93,34	93,46	93,57	93,69	93,80	93,92	94,03
82	94,15	94,26	94,38	94,49	94,61	94,72	94,83	94,95	95,06	95,18
83	95,29	95,41	95,52	95,64	95,75	95,87	95,98	96,10	96,21	96,33
84	96,44	96,56	96,67	96,79	96,90	97,02	97,13	97,25	97,36	97,48
85	97,59	97,71	97,82	97,93	98,05	98,16	98,28	98,39	98,51	98,62
86	98,74	98,85	98,97	99,08	99,20	99,31	99,43	99,54	99,66	99,77
87	99,89	100,00	100,12	100,23	100,35	100,46	100,58	100,69	100,80	100,97
88	101,03	101,15	101,26	101,38	101,49	101,61	101,72	101,84	101,95	102,02
89	102,18	102,30	102,41	102,53	102,64	102,76	102,87	102,99	103,10	103,22
90	103,33	103,45	103,56	103,68	103,79	103,90	104,02	104,13	104,25	104,36
91	104,48	104,59	104,71	104,82	104,94	105,05	105,17	105,28	105,40	105,51
92	105,63	105,74	105,86	105,97	106,09	106,20	106,32	106,43	106,55	106,66
93	106,78	106,89	107,00	107,12	107,23	107,35	107,46	107,58	107,69	107,81
94	107,92	108,04	108,15	108,27	108,38	108,50	108,61	108,73	108,84	108,96
95	109,07	109,19	109,30	109,42	109,53	109,65	109,76	109,88	109,99	110,10
96	110,22	110,33	110,45	110,56	110,68	110,79	110,91	111,02	111,14	111,25
97	111,37	111,48	111,60	111,71	111,83	111,94	112,06	112,17	112,29	112,40
98	112,52	112,63	112,75	112,86	112,98	113,09	113,20	113,32	113,43	113,55
99	113,66	113,78	113,89	114,01	114,12	114,24	114,35	114,47	114,59	114,70
100	114,81	114,93	115,04	115,16	115,27	115,39	115,50	115,62	115,73	115,85

Table 49

Conversion of Milligrams of SO_3 into Milligrams of SO_4^{2-}

Whole mg of SO_3	Tenths of a mg of SO_3									
	0	1	2	3	4	5	6	7	8	9
0	—	0,1	0,2	0,4	0,5	0,6	0,7	0,8	1,0	1,1
1	1,20	1,32	1,44	1,56	1,68	1,80	1,92	2,04	2,16	2,28
2	2,40	2,52	2,64	2,76	2,88	3,00	3,12	3,24	3,36	3,48
3	3,60	3,72	3,84	3,96	4,08	4,20	4,32	4,44	4,56	4,68
4	4,80	4,92	5,04	5,16	5,28	5,40	5,52	5,64	5,76	5,88
5	6,00	6,12	6,24	6,36	6,48	6,60	6,72	6,84	6,96	7,08
6	7,20	7,32	7,44	7,56	7,68	7,80	7,92	8,04	8,16	8,28
7	8,40	8,52	8,64	8,76	8,88	9,00	9,12	9,24	9,36	9,48
8	9,60	9,72	9,84	9,96	10,08	10,20	10,32	10,44	10,56	10,68
9	10,80	10,92	11,04	11,16	11,28	11,40	11,52	11,64	11,76	11,88
10	12,00	12,12	12,24	12,36	12,48	12,60	12,72	12,84	12,96	13,08
11	13,20	13,32	13,44	13,56	13,68	13,80	13,92	14,04	14,16	14,28
12	14,40	14,52	14,64	14,76	14,88	15,00	15,12	15,24	15,36	15,48
13	15,60	15,72	15,84	15,96	16,08	16,20	16,32	16,44	16,56	16,68
14	16,80	16,92	17,04	17,16	17,28	17,40	17,52	17,64	17,76	17,88
15	18,00	18,12	18,24	18,36	18,48	18,60	18,72	18,84	18,96	19,08
16	19,20	19,32	19,44	19,56	19,68	19,80	19,92	20,04	20,16	20,28
17	20,40	20,52	20,64	20,76	20,88	21,00	21,12	21,24	21,36	21,48
18	21,60	21,72	21,84	21,96	22,08	22,20	22,32	22,44	22,56	22,68
19	22,80	22,92	23,04	23,16	23,28	23,40	23,52	23,64	23,76	23,88
20	24,00	24,12	24,24	24,36	24,48	24,60	24,72	24,84	24,96	25,08
21	25,20	25,32	25,44	25,56	25,68	25,80	25,92	26,04	26,16	26,28
22	26,40	26,52	26,64	26,76	26,88	27,00	27,12	27,24	27,36	27,48
23	27,60	27,72	27,84	27,96	28,08	28,20	28,32	28,44	28,56	28,68
24	28,80	28,92	29,04	29,16	29,28	29,40	29,52	29,64	29,76	29,88
25	30,00	30,12	30,24	30,36	30,48	30,60	30,72	30,84	30,96	31,08
26	31,20	31,32	31,44	31,56	31,68	31,80	31,92	32,04	32,16	32,28
27	32,40	32,52	32,64	32,76	32,88	33,00	33,12	33,24	33,36	33,48
28	33,60	33,72	33,84	33,96	34,08	34,20	34,32	34,44	34,56	34,68
29	34,80	34,92	35,04	35,16	35,28	35,40	35,52	35,64	35,76	35,88
30	36,00	36,12	36,24	36,36	36,47	36,59	36,71	36,83	36,95	37,07
31	37,19	37,31	37,43	37,55	37,67	37,79	37,91	38,03	38,15	38,27
32	38,39	38,51	38,63	38,75	38,87	38,99	39,11	39,23	39,35	39,47
33	39,59	39,71	39,83	39,95	40,07	40,19	40,31	40,43	40,55	40,67
34	40,79	40,91	41,03	41,15	41,27	41,39	41,51	41,63	41,75	41,87
35	41,99	42,11	42,23	42,35	42,47	42,59	42,71	42,83	42,95	43,07
36	43,19	43,31	43,43	43,55	43,67	43,79	43,91	44,03	44,15	44,27
37	44,39	44,51	44,63	44,75	44,87	44,99	45,11	45,23	45,35	45,47
38	45,59	45,71	45,83	45,95	46,07	46,19	46,31	46,43	46,55	46,67
39	46,79	46,91	47,03	47,15	47,27	47,39	47,51	47,63	47,75	47,87
40	47,99	48,11	48,23	48,35	48,47	48,59	48,71	48,83	48,95	49,07
41	49,19	49,31	49,43	49,55	49,67	49,79	49,91	50,03	50,15	50,27
42	50,39	50,51	50,63	50,75	50,87	50,99	51,11	51,23	51,35	51,47
43	51,59	51,71	51,83	51,95	52,07	52,19	52,31	52,43	52,55	52,67
44	52,79	52,91	53,03	53,15	53,27	53,39	53,51	53,63	53,75	53,87
45	53,99	54,11	54,23	54,35	54,47	54,59	54,71	54,83	54,95	55,07
46	55,19	55,31	55,43	55,55	55,67	55,79	55,91	56,03	56,15	56,27
47	56,39	56,51	56,63	56,75	56,87	56,99	57,11	57,23	57,35	57,47
48	57,59	57,71	57,83	57,95	58,07	58,19	58,31	58,43	58,55	58,67
49	58,79	58,91	59,03	59,15	59,27	59,39	59,51	59,63	59,75	59,87
50	59,99	60,11	60,23	60,35	60,47	60,59	60,71	60,83	60,95	61,07

Whole mg of SO₃	Tenths of a mg of SO₃									
	0	1	2	3	4	5	6	7	8	9
51	61,19	61,31	61,43	61,55	61,67	61,79	61,91	62,03	62,15	62,27
52	62,39	62,51	62,63	62,75	62,87	62,99	63,11	63,23	63,35	63,47
53	63,59	63,71	63,83	63,95	64,07	64,19	64,31	64,43	64,55	64,67
54	64,79	64,91	65,03	65,15	65,27	65,39	65,51	65,63	65,75	65,87
55	65,99	66,11	66,23	66,35	66,47	66,59	66,71	66,83	66,95	67,07
56	67,19	67,31	67,43	67,55	67,67	67,79	67,91	68,03	68,15	68,27
57	68,39	68,51	68,63	68,75	68,87	68,99	69,11	69,23	69,35	69,47
58	69,59	69,71	69,83	69,95	70,07	70,19	70,31	70,43	70,55	70,67
59	70,79	70,91	71,03	71,15	71,27	71,39	71,51	71,63	71,75	71,87
60	71,99	72,11	72,23	72,35	72,47	72,59	72,71	72,83	72,95	73,07
61	73,19	73,31	73,43	73,55	73,67	73,79	73,91	74,03	74,15	74,27
62	74,39	74,51	74,63	74,75	74,87	74,99	75,11	75,23	75,35	75,47
63	75,59	75,71	75,83	75,95	76,07	76,19	76,31	76,43	76,55	76,67
64	76,79	76,91	77,03	77,15	77,27	77,39	77,51	77,63	77,75	77,87
65	77,99	78,11	78,23	78,35	78,47	78,59	78,71	78,83	78,95	79,07
66	79,19	79,31	79,43	79,55	79,67	79,79	79,91	80,03	80,15	80,27
67	80,39	80,51	80,63	80,75	80,87	80,99	81,11	81,23	81,35	81,47
68	81,59	81,71	81,83	81,95	82,07	82,19	82,31	82,43	82,55	82,67
69	82,79	82,91	83,03	83,15	83,27	83,39	83,51	83,63	83,75	83,87
70	83,99	84,11	84,23	84,35	84,47	84,59	84,71	84,83	84,95	85,07
71	85,19	85,31	85,43	85,55	85,67	85,79	85,91	86,03	86,15	86,27
72	86,39	86,51	86,63	86,75	86,87	86,99	87,11	87,23	87,35	87,47
73	87,59	87,71	87,83	87,95	88,07	88,19	88,31	88,43	88,55	88,67
74	88,79	88,91	89,03	89,15	89,27	89,39	89,51	89,63	89,75	89,87
75	89,99	90,11	90,23	90,35	90,47	90,59	90,71	90,83	90,95	91,07
76	91,19	91,31	91,43	91,55	91,67	91,79	91,91	92,03	92,15	92,27
77	92,39	92,51	92,63	92,75	92,87	92,99	93,11	93,23	93,35	93,47
78	93,59	93,71	93,83	93,95	94,07	94,19	94,31	94,43	94,55	94,67
79	94,79	94,91	95,03	95,15	95,27	95,39	95,51	95,63	95,75	95,87
80	95,99	96,11	96,23	96,35	96,47	96,59	96,71	96,83	96,95	97,07
81	97,19	97,31	97,43	97,55	97,67	97,79	97,91	98,03	98,15	98,27
82	98,39	98,51	98,63	98,75	98,87	98,99	99,11	99,23	99,35	99,47
83	99,59	99,71	99,83	99,95	100,07	100,19	100,31	100,43	100,55	100,67
84	100,79	100,91	101,03	101,15	101,27	101,39	101,51	101,63	101,75	101,87
85	101,99	102,11	102,23	102,35	102,47	102,59	102,71	102,83	102,95	103,07
86	103,19	103,31	103,43	103,55	103,67	103,79	103,91	104,03	104,15	104,27
87	104,39	104,51	104,63	104,75	104,87	104,99	105,11	105,23	105,35	105,47
88	105,59	105,71	105,83	105,95	106,07	106,19	106,31	106,43	106,55	106,67
89	106,79	106,91	107,03	107,15	107,27	107,39	107,51	107,63	107,75	107,87
90	107,99	108,11	108,23	108,35	108,47	108,59	108,71	108,83	108,95	109,07
91	109,18	109,30	109,42	109,54	109,66	109,78	109,90	110,02	110,14	110,26
92	110,38	110,50	110,62	110,74	110,86	110,98	111,10	111,22	111,34	111,46
93	111,58	111,70	111,82	111,94	112,06	112,18	112,30	112,42	112,54	112,66
94	112,78	112,90	113,02	113,14	113,26	113,38	113,50	113,62	113,74	113,86
95	113,98	114,10	114,22	114,34	114,46	114,58	114,70	114,82	114,94	115,06
96	115,18	115,30	115,42	115,54	115,66	115,78	115,90	116,02	116,14	116,26
97	116,38	116,50	116,62	116,74	116,86	116,98	117,10	117,22	117,34	117,46
98	117,58	117,70	117,82	117,94	118,06	118,18	118,30	118,42	118,54	118,66
99	118,78	118,90	119,02	119,14	119,26	119,38	119,50	119,62	119,74	119,86
100	119,98	120,10	120,22	120,34	120,46	120,58	120,70	120,82	120,94	121,06

mg of SO₃	1000	2000	3000	4000	5000	6000	7000	8000	9000	10 000
mg of SO₄²⁻	1199,8	2399,7	3599,5	4799,3	5999,2	7199,0	8398,8	9598,7	10798,5	11 938

Table 50

Conversion of Milligrams of Bicarbonate CO_2 into Milligrams of HCO_3^-

Whole mg of CO_2	Tenths of a mg of CO_2									
	0	1	2	3	4	5	6	7	8	9
0	—	0,1	0,3	0,4	0,6	0,7	0,8	1,0	1,1	1,2
1	1,39	1,52	1,66	1,80	1,94	2,08	2,22	2,36	2,50	2,63
2	2,77	2,91	3,06	3,19	3,33	3,47	3,60	3,74	3,88	4,02
3	4,16	4,30	4,44	4,57	4,71	4,85	4,99	5,13	5,27	5,41
4	5,55	5,68	5,82	5,96	6,10	6,24	6,38	6,52	6,66	6,79
5	6,93	7,07	7,21	7,35	7,49	7,63	7,76	7,90	8,04	8,18
6	8,32	8,46	8,60	8,73	8,87	9,01	9,15	9,29	9,43	9,57
7	9,71	9,84	9,98	10,12	10,26	10,40	10,54	10,68	10,81	10,95
8	11,09	11,23	11,37	11,51	11,65	11,78	11,92	12,06	12,20	12,34
9	12,48	12,62	12,76	12,89	13,03	13,17	13,31	13,45	13,59	13,73
10	13,86	14,00	14,14	14,28	14,42	14,56	14,70	14,84	14,97	15,11
11	15,25	15,39	15,53	15,67	15,81	15,94	16,08	16,22	16,36	16,50
12	16,64	16,78	16,91	17,05	17,19	17,33	17,47	17,61	17,75	17,89
13	18,02	18,16	18,30	18,44	18,58	18,72	18,86	18,99	19,13	19,27
14	19,41	19,55	19,69	19,83	19,96	20,10	20,24	20,38	20,52	20,66
15	20,80	20,94	21,07	21,21	21,35	21,49	21,63	21,77	21,91	22,04
16	22,18	22,32	22,46	22,60	22,74	22,88	23,02	23,15	23,29	23,43
17	23,57	23,71	23,85	23,99	24,12	24,26	24,40	24,54	24,68	24,82
18	24,96	25,09	25,23	25,37	25,51	25,65	25,79	25,93	26,07	26,20
19	26,34	26,48	26,62	26,76	26,90	27,04	27,17	27,31	27,45	27,59
20	27,73	27,87	28,01	28,15	28,28	28,42	28,56	28,70	28,84	28,98
21	29,12	29,25	29,39	29,53	29,67	29,81	29,95	30,09	30,22	30,36
22	30,50	30,64	30,78	30,92	31,06	31,20	31,33	31,47	31,61	31,75
23	31,89	32,03	32,17	32,30	32,44	32,58	32,72	32,86	33,00	33,14
24	33,27	33,41	33,55	33,69	33,83	33,97	34,11	34,25	34,38	34,52
25	34,66	34,80	34,94	35,08	35,22	35,35	35,49	35,63	35,77	35,91
26	36,05	36,19	36,32	36,46	36,60	36,74	36,88	37,02	37,16	37,30
27	37,43	37,57	37,71	37,85	37,99	38,13	38,27	38,40	38,54	38,68
28	38,82	38,96	39,10	39,24	39,38	39,51	39,65	39,79	39,93	40,07
29	40,21	40,35	40,48	40,62	40,76	40,90	41,04	41,18	41,32	41,45
30	41,59	41,73	41,87	42,01	42,15	42,29	42,43	42,56	42,70	42,84
31	42,98	43,12	43,26	43,40	43,53	43,67	43,81	43,95	44,09	44,23
32	44,37	44,51	44,64	44,78	44,92	45,06	45,20	45,34	45,48	45,61
33	45,75	45,89	46,03	46,17	46,31	46,45	46,58	46,72	46,86	47,00
34	47,14	47,28	47,42	47,56	47,69	47,83	47,97	48,11	48,25	48,39
35	48,53	48,66	48,80	48,94	49,08	49,22	49,36	49,50	49,63	49,77
36	49,91	50,05	50,19	50,33	50,47	50,61	50,74	50,88	51,02	51,16
37	51,30	51,44	51,58	51,71	51,85	51,99	52,13	52,27	52,41	52,55
38	52,69	52,82	52,96	53,10	53,24	53,38	53,52	53,66	53,79	53,93
39	54,07	54,21	54,35	54,49	54,63	54,76	54,90	55,04	55,18	55,32
40	55,46	55,60	55,74	55,87	56,01	56,15	56,29	56,43	56,57	56,71
41	56,84	56,98	57,12	57,26	57,40	57,54	57,68	57,81	57,95	58,09
42	58,23	58,37	58,51	58,65	58,79	58,92	59,06	59,20	59,34	59,48
43	59,62	59,76	59,89	60,03	60,17	60,31	60,45	60,59	60,73	60,87
44	61,00	61,14	61,28	61,42	61,56	61,70	61,84	61,97	62,11	62,25
45	62,39	62,53	62,67	62,81	62,94	63,08	63,22	63,36	63,50	63,64
46	63,78	63,92	64,05	64,19	64,33	64,47	64,61	64,75	64,89	65,02
47	65,16	65,30	65,44	65,58	65,72	65,86	66,00	66,13	66,27	66,41
48	66,55	66,69	66,83	66,97	67,10	67,24	67,38	67,52	67,66	67,80
49	67,94	68,07	68,21	68,35	68,49	68,63	68,77	68,91	69,05	69,18
50	69,32	69,46	69,60	69,74	69,88	70,02	70,15	70,29	70,43	70,57

Whole mg of CO$_2$	Tenths of a mg of CO$_2$									
	0	1	2	3	4	5	6	7	8	9
51	70,71	70,85	70,99	71,12	71,26	71,40	71,54	71,68	71,82	71,96
52	72,10	72,23	72,37	72,51	72,65	72,79	72,93	73,07	73,20	73,34
53	73,48	73,62	73,76	73,90	74,04	74,18	74,31	74,45	74,59	74,73
54	74,87	75,01	75,15	75,28	75,42	75,56	75,70	75,84	75,98	76,12
55	76,25	76,39	76,53	76,67	76,81	76,95	77,09	77,23	77,36	77,50
56	77,64	77,78	77,92	78,06	78,20	78,33	78,47	78,61	78,75	78,89
57	79,03	79,17	79,30	79,44	79,58	79,72	79,86	80,00	80,14	80,28
58	80,41	80,55	80,69	80,83	80,97	81,11	81,25	81,38	81,52	81,66
59	81,80	81,94	82,08	82,22	82,36	82,49	82,63	82,77	82,91	83,05
60	83,19	83,33	83,46	83,60	83,74	83,88	84,02	84,16	84,30	84,43
61	84,57	84,71	84,85	84,99	85,13	85,27	85,41	85,54	85,68	85,82
62	85,96	86,10	86,24	86,38	86,51	86,65	86,79	86,93	87,07	87.21
63	87,35	87,48	87,62	87,76	87,90	88,04	88,18	88,32	88,46	88,59
64	88,73	88,87	89,01	89,15	89,29	89,43	89,56	89,70	89,84	89,98
65	90,12	90,26	90,40	90,54	90,67	90,81	90,95	91,09	91,23	91,37
66	91,51	91,64	91,78	91,92	92,06	92,20	92,34	92,48	92,61	92,75
67	92,89	93,03	93,17	93,31	93,45	93,59	93,72	93,86	94,00	94,14
68	94,28	94,42	94,56	94,69	94,83	94,97	95,11	95,25	95,39	95,53
69	95,66	95,80	95,94	96,08	96,22	96,36	96,50	96,64	96,77	96,91
70	97,05	97,19	97,33	97,47	97,61	97,74	97,88	98,02	98,16	98,30
71	98,44	98,58	98,72	98,85	98,99	99,13	99,27	99,41	99,55	99,69
72	99,82	99,96	100,10	100,24	100,38	100,52	100,66	100,79	100,93	101,07
73	101,21	101,35	101,49	101,63	101,77	101,90	102,04	102,18	102,32	102,46
74	102,60	102,74	102,87	103,01	103,15	103,29	103,43	103,57	103,71	103,85
75	103,98	104,12	104,26	104,40	104,54	104,68	104,82	104,95	105,09	105,23
76	105,37	105,51	105,65	105,79	105,92	106,06	106,20	106,34	106,48	106,62
77	106,76	106,90	107,03	107,17	107,31	107,45	107,59	107,73	107,87	108,00
78	108,14	108,28	108,42	108,56	108,70	108,84	108,97	109,11	109,25	109,39
79	109,53	109,67	109,81	109,95	110,08	110,22	110,36	110,50	110,64	110,78
80	110,92	111,05	111,19	111,33	111,47	111,61	111,75	111,89	112,03	112,16
81	112,30	112,44	112,58	112,72	112,86	113,00	113,13	113,27	113,41	113,65
82	113,69	113,83	113,97	114,10	114,24	114,38	114,52	114,66	114,80	114,94
83	115,08	115,21	115,35	115,49	115,63	115,77	115,91	116,05	116,18	116,32
84	116,46	116,60	116,74	116,88	117,02	117,15	117,29	117,43	117,57	117,71
85	117,85	117,99	118,13	118,26	118,40	118,54	118,68	118,82	118,96	119,10
86	119,23	119,37	119,51	119,65	119,79	119,93	120,07	120,21	120,34	120,48
87	120,62	120,76	120,90	121,04	121,18	121,31	121,45	121,59	121,73	121,87
88	122,01	122,15	122,28	122,42	122,56	122,70	122,84	122,98	123,12	123,26
89	123,39	123,53	123,67	123,81	123,95	124,09	124,23	124,36	124,50	124,64
90	124,78	124,92	125,06	125,20	125,33	125,47	125,61	125,75	125,89	126,03
91	126,17	126,31	126,44	126,58	126,72	126,86	127,00	127,14	127,28	127,41
92	127,55	127,69	127,83	127,97	128,11	128,25	128,39	128,52	128,66	128,80
93	128,94	129,08	129,22	129,36	129,49	129,63	129,77	129,91	130,05	130,19
94	130,33	130,46	130,60	130,74	130,88	131,02	131,16	131,30	131,44	131,57
95	131,71	131,85	131,99	132,13	132,27	132,41	132,54	132,68	132,82	132,96
96	133,10	133,24	133,38	133,52	133,65	133,79	133,93	134,07	134,21	134,35
97	134,49	134,62	134,76	134,90	135,04	135,18	135,32	135,46	135,59	135,73
98	135,87	136,01	136,15	136,29	136,43	136,57	136,70	136,84	136,98	137,12
99	137,26	137,40	137,54	137,67	137,81	137,95	138,09	138,23	138,37	138,51
100	138,64	138,78	138,92	139,06	139,20	139,34	139,48	139,62	139,75	139,89

Table 51

Conversion of Milligrams of Partly Bound CO_2 into Milligrams of HCO_3^-

Whole mg of CO_2	Tenths of a mg of CO_2									
	0	1	2	3	4	5	6	7	8	9
0	—	0,3	0,6	0,8	1,1	1,6	1,7	1,9	2,2	2,5
1	2,77	3,05	3,33	3,60	3,88	4,16	4,44	4,71	4,99	5,27
2	5,55	5,82	6,10	6,38	6,66	6,93	7,21	7,49	7,76	8,04
3	8,32	8,60	8,87	9,15	9,43	9 70	9,98	10,26	10,54	10,81
4	11,09	11,37	11,65	11,92	12,20	12,48	12,75	13,03	13,31	13,59
5	13,86	14,14	14,42	14,70	14,97	15,25	15,53	15,81	16,08	16,36
6	16,64	16,91	17,19	17,47	17,75	18,02	18,30	18,58	18,85	19,13
7	19,41	19,69	19,96	20,24	20,52	20,80	21,07	21,35	21,63	21,91
8	22,18	22,46	22,74	23,01	23,29	23,57	23,85	24,12	24,40	24,68
9	24,96	25,23	25,51	25,79	26,06	26,34	26,62	26,90	27,17	27,45
10	27,73	28,01	28,28	28,56	28,84	29,12	29,39	29,67	29,95	30,22
11	30,50	30,78	31,06	31,33	31,61	31,89	32,17	32,44	32,72	33,00
12	33,27	33,55	33,83	34,11	34,38	34,66	34,94	35,22	35,49	35,77
13	36,05	36,32	36,60	36,88	37,16	37,43	37,71	37,99	38,27	38,54
14	38,82	39,10	39,38	39,65	39,93	40,21	40,48	40,76	41,04	41,32
15	41,59	41,87	42,15	42,43	42,70	42,98	43,26	43,53	43,81	44,09
16	44,37	44,64	44,92	45,20	45,48	45,75	46,03	46,31	46,58	46,86
17	47,14	47,42	47,69	47,97	48,25	48,53	48,80	49,08	49,36	49,63
18	49,91	50,19	50,47	50,74	51,02	51,30	51,58	51,85	52,13	52,41
19	52,69	52,96	53,24	53,52	53,79	54,07	54,35	54,63	54,90	55,18
20	55,46	55,74	56,01	56,29	56,57	56,84	57,12	57,40	57,68	57,95
21	58,23	58,51	58,79	59,06	59,34	59,62	59,89	60,17	60,45	60,73
22	61,00	61,28	61,56	61,84	62,11	62,39	62,67	62,94	63,22	63,50
23	63,78	64,05	64,33	64,61	64 89	65,16	65,44	65,72	66,00	66,27
24	66,55	66,83	67,10	67,38	67,66	67,94	68,21	68,49	68,77	69,05
25	69,32	69,60	69,88	70,15	70,43	70,71	70,99	71,26	71,54	71,82
26	72,10	72,37	72,65	72,93	73,20	73,48	73,76	74,04	74,31	74,59
27	74,87	75,15	75,42	75,70	75,98	76,25	76,53	76,81	77,09	77,36
28	77,64	77,92	78,20	78,47	78,75	79,03	79,30	79,58	79,86	80,14
29	80,41	80,69	80,97	81,25	81,52	81,80	82,08	82,36	82,63	82,91
30	83,19	83,46	83,74	84,02	84,30	84,57	84,85	85,13	85,41	85,68
31	85,96	86,24	86,51	86,79	87,07	87,35	87,62	87,90	88,18	88,46
32	88,73	89,01	89,29	89,56	89,84	90,12	90,40	90,67	90,95	91,23
33	91,51	91,78	92,06	92,34	92,61	92,89	93,17	93,45	93,72	94,00
34	94,28	94,56	94,83	95,11	95,39	95,66	95,94	96,22	96,50	96,77
35	97,05	97,33	97,61	97,88	98,16	98,44	98,72	98,99	99,27	99,55
36	99,82	100,10	100,38	100,66	100,93	101,21	101,49	101,77	102,04	102,32
37	102,60	102,87	103,15	103,43	103,71	103,98	104,26	104,54	104,82	105,09
38	105,37	105,65	105,92	106,20	106,48	106,76	107,03	107,31	107,59	107,87
39	108,14	108,42	108,70	108,97	109,25	109,53	109,81	110,08	110,36	110,64
40	110,92	111,19	111,47	111,75	112,03	112,30	112,58	112,86	113,13	113,41
41	113,69	113,97	114,24	114,52	114,80	115,08	115,35	115,63	115,91	116,18
42	116,46	116,74	117,02	117,29	117,57	117,85	118,13	118,40	118,68	118,96
43	119,23	119,51	119,79	120,07	120,34	120,62	120,90	121,18	121,45	121,73
44	122,01	122,28	122,56	122,84	123,12	123,39	123,67	123,95	124,23	124,50
45	124,78	125,06	125,33	125,61	125,89	126,17	126,44	126,72	127,00	127,28
46	125,55	127,83	128,11	128,39	128,66	128,94	129,22	129,49	129,77	130,05
47	130,33	130,60	130,88	131,16	131,44	131,71	131,99	132,27	132,54	132,82
48	133,10	133,38	133,65	133,93	134,21	134,49	134,76	135,04	135,32	135,59
49	135,87	136,15	136,43	136,70	136,98	137,26	137,54	137,81	138,09	138,37
50	138,64	138,92	139,20	139,48	139,75	140,03	140,31	140,59	140,86	141,14

Whole mg of CO_2	Tenths of a mg of CO_2									
	0	1	2	3	4	5	6	7	8	9
51	141,42	141,70	141,97	142,25	142,53	142,80	143,08	143,36	143,64	143,91
52	144,19	144,47	144,75	145,02	145,30	145,58	145,85	146,13	146,41	146,69
53	146,96	147,24	147,52	147,80	148,07	148,35	148,63	148,90	149,18	149,46
54	149,74	150,01	150,29	150,57	150,85	151,12	151,40	151,68	151,95	152,23
55	152,51	152,79	153,06	153,34	153,62	153,90	154,17	154,45	154,73	155,00
56	155,28	155,56	155,84	156,11	156,39	156,67	156,95	157,22	157,50	157,78
57	158,06	158,33	158,61	158,89	159,16	159,44	159,72	160,00	160,27	160,55
58	160,83	161,11	161,38	161,66	161,94	162,21	162,49	162,77	163,05	163,32
59	163,60	163,88	164,16	164,43	164,71	164,99	165,26	165,54	165,82	166,10
60	166,37	166,65	166,93	167,21	167,48	167,76	168,04	168,31	168,59	168,87
61	169,15	169,42	169,70	169,98	170,26	170,53	170,81	171,09	171,37	171,64
62	171,92	172,20	172,47	172,75	173,03	173,31	173,58	173,86	174,14	174,42
63	174,69	174,97	175,25	175,52	175,80	176,08	176,36	176,63	176,91	177,19
64	177,47	177,74	178,02	178,30	178,57	178,85	179,13	179,41	179,68	179,96
65	180,24	180,52	180,79	181,07	181,35	181,62	181,90	182,18	182,46	182,73
66	183,01	183,29	183,57	183,84	184,12	184,40	184,67	184,95	185,23	185,51
67	185,78	186,06	186,34	186,62	186,89	187,17	187,45	187,73	188,00	188,28
68	188,56	188,83	189,11	189,39	189,67	189,94	190,22	190,50	190,78	191,05
69	191,33	191,61	191,88	192,16	192,44	192,72	192,99	193,27	193,55	193,83
70	194,10	194,38	194,66	194,93	195,21	195,49	195,77	196,04	196,32	196,60
71	196,88	197,15	197,43	197,71	197,98	198,26	198,54	198,82	199,09	199,37
72	199,65	199,93	200,20	200,48	200,76	201,04	201,31	201,59	201,87	202,14
73	202,42	202,70	202,98	203,25	203,53	203,81	204,09	204,36	204,64	204,92
74	205,19	205,47	205,75	206,03	206,30	206,58	206,86	207,14	207,41	207,69
75	207,97	208,24	208,52	208,80	209,08	209,35	209,63	209,91	210,19	210,46
76	210,74	211,02	211,29	211,57	211,85	212,13	212,40	212,68	212,96	213,24
77	213,51	213,79	214,07	214,35	214,62	214,90	215,18	215,45	215,73	216,01
78	216,29	216,56	216,84	217,12	217,40	217,67	217,95	218,23	218,50	218,78
79	219,06	219,34	219,61	219,89	220,17	220,45	220,72	221,00	221,28	221,55
80	221,83	222,11	222,39	222,66	222,94	223,22	223,50	223,77	224,05	224,33
81	224,60	224,88	225,16	225,44	225,71	225,99	226,27	226,55	226,82	227,10
82	227,38	227,65	227,93	228,21	228,49	228,76	229,04	229,32	229,60	229,87
83	230,15	230,43	230,71	230,98	231,26	231,54	231,81	232,09	232,37	232,65
84	232,92	233,20	233,48	233,76	234,03	234,31	234,59	234,86	235,14	235,42
85	235,70	235,97	236,25	236,53	236,81	237,08	237,36	237,64	237,91	238,19
86	238,47	238,75	239,02	239,30	239,58	239,86	240,13	240,41	240,69	240,96
87	241,24	241,52	241,80	242,07	242,35	242,63	242,91	243,18	243,46	243,74
88	244,01	244,29	244,57	244,85	245,12	245,40	245,68	245,96	246,23	246,51
89	246,79	247,07	247,34	247,62	247,90	248,17	248,45	248,73	249,01	249,28
90	249,56	249,84	250,12	250,39	250,67	250,95	251,22	251,50	251,78	252,06
91	252,33	252,61	252,89	253,17	253,44	253,72	254,00	254,27	254,55	254,83
92	255,11	255,38	255,66	255,94	256,22	256,49	256,77	257,05	257,32	257,60
93	257,88	258,16	258,43	258,71	258,99	259,27	259,54	259,82	260,10	260,38
94	260,65	260,93	261,21	261,48	261,76	262,04	262,32	262,59	262,87	263,15
95	263,43	263,70	263,98	264,26	264,53	264,81	265,09	265,37	265,64	265,92
96	266,20	266,48	266,75	267,03	267,31	267,58	267,86	268,14	268,42	268,69
97	268,97	269,25	269,53	269,80	270,08	270,36	270,63	270,91	271,19	271,47
98	271,74	272,02	272,30	272,58	272,85	273,13	273,41	273,68	273,96	274,24
99	274,52	274,79	275,07	275,35	275,63	275,90	276,18	276,46	276,74	277,01
100	277,29	277,57	277,84	278,12	278,40	278,68	278,95	279,23	279,51	279,79

Table 52

Conversion of Milligrams of Carbonate CO_2 into Milligrams of CO_3^{2-}

Whole mg of CO_2	Tenths of a mg of CO_2									
	0	1	2	3	4	5	6	7	8	9
0	—	0,1	0,3	0,4	0,5	0,7	0,8	1,0	1,1	1,2
1	1,36	1,50	1,64	1,77	1,91	2,05	2,18	2,32	2,45	2,59
2	2,73	2,86	3,00	3,14	3,27	3,41	3,55	3,68	3,82	3,95
3	4,09	4,23	4,36	4,50	4,64	4,77	4,91	5,05	5,18	5,32
4	5,45	5,59	5,73	5,86	6,00	6,14	6,27	6,41	6,55	6,68
5	6,82	6,95	7,09	7,23	7,36	7,50	7,64	7,77	7,91	8,04
6	8,18	8,32	8,45	8,59	8,73	8,86	9,00	9,14	9,27	9,41
7	9,54	9,68	9,82	9,95	10,09	10,23	10,36	10,50	10,64	10,77
8	10,91	11,04	11,18	11,32	11,45	11,59	11,73	11,86	12,00	12,14
9	12,27	12,41	12,54	12,68	12,82	12,95	13,09	13,23	13,36	13,50
10	13,64	13,77	13,91	14,04	14,18	14,32	14,45	14,59	14,73	14,86
11	15,00	15,14	15,27	15,41	15,54	15,68	15,82	15,95	16,09	16,23
12	16,36	16,50	16,64	16,77	16,91	17,04	17,18	17,32	17,45	17,59
13	17,73	17,86	18,00	18,14	18,27	18,41	18,54	18,68	18,82	18,95
14	19,09	19,23	19,36	19,50	19,64	19,77	19,91	20,04	20,18	20,32
15	20,45	20,59	20,73	20,86	21,00	21,13	21,27	21,41	21,54	21,68
16	21,82	21,95	22,09	22,23	22,36	22,50	22,63	22,77	22,91	23,04
17	23,18	23,32	23,45	23,59	23,73	23,86	24,00	24,13	24,27	24,41
18	24,54	24,68	24,82	24,95	25,09	25,23	25,36	25,50	25,63	25,77
19	25,91	26,04	26,18	26,32	26,45	26,59	26,73	26,86	27,00	27,13
20	27,27	27,41	27,54	27,68	27,82	27,95	28,09	28,23	28,36	28,50
21	28,63	28,77	28,91	29,04	29,18	29,32	29,45	29,59	29,73	29,86
22	30,00	30,13	30,27	30,41	30,54	30,68	30,82	30,95	31,09	31,23
23	31,36	31,50	31,63	31,77	31,91	32,04	32,18	32,32	32,45	32,59
24	32,73	32,86	33,00	33,13	33,27	33,41	33,54	33,68	33,82	33,95
25	34,09	34,22	34,36	34,50	34,63	34,77	34,91	35,04	35,18	35,32
26	35,45	35,59	35,72	35,86	36,00	36,13	36,27	36,41	36,54	36,68
27	36,82	36,95	37,09	37,22	37,36	37,50	37,63	37,77	37,91	38,04
28	38,18	38,32	38,45	38,59	38,72	38,86	39,00	39,13	39,27	39,41
29	39,54	39,68	39,82	39,95	40,09	40,22	40,36	40,50	40,63	40,77
30	40,91	41,04	41,18	41,32	41,45	41,59	41,72	41,86	42,00	42,13
31	42,27	42,41	42,54	42,68	42,82	42,95	43,09	43,22	43,36	43,50
32	43,63	43,77	43,91	44,04	44,18	44,32	44,45	44,59	44,72	44,86
33	45,00	45,13	45,27	45,41	45,54	45,68	45,82	45,95	46,09	46,22
34	46,36	46,50	46,63	46,77	46,91	47,04	47,18	47,32	47,45	47,59
35	47,72	47,86	48,00	48,13	48,27	48,41	48,54	48,68	48,81	48,95
36	49,09	49,22	49,36	49,50	49,63	49,77	49,91	50,04	50,18	50,31
37	50,45	50,59	50,72	50,86	51,00	51,13	51,27	51,41	51,54	51,68
38	51,81	51,95	52,09	52,22	52,36	52,50	52,63	52,77	52,91	53,04
39	53,18	53,31	53,45	53,59	53,72	53,86	54,00	54,13	54,27	54,41
40	54,54	54,68	54,81	54,95	55,09	55,22	55,36	55,50	55,63	55,77
41	55,91	56,04	56,18	56,31	56,45	56,59	56,72	56,86	57,00	57,13
42	57,27	57,41	57,54	57,68	57,81	57,95	58,09	58,22	58,36	58,50
43	58,63	58,77	58,91	59,04	59,18	59,31	59,45	59,59	59,72	59,86
44	60,00	60,13	60,27	60,41	60,54	60,68	60,81	60,95	61,09	61,22
45	61,36	61,50	61,63	61,77	61,90	62,04	62,18	62,31	62,45	62,59
46	62,72	62,86	63,00	63,13	63,27	63,40	63,54	63,68	63,81	63,95
47	64,09	64,22	64,36	64,50	64,63	64,77	64,90	65,04	65,18	65,31
48	65,45	65,59	65,72	65,86	66,00	66,13	66,27	66,40	66,54	66,68
49	66,81	66,95	67,09	67,22	67,36	67,50	67,63	67,77	67,90	68,04
50	68,18	68,31	68,45	68,59	68,72	68,86	69,00	69,13	69,27	69,40

VI. Tables for Converting Nitrogen into Nitrogen-Containing Ions

Table 53

Conversion of Milligrams of N into Milligrams of NH_4^+

Whole mg of N	Tenths of a mg of N									
	0	1	2	3	4	5	6	7	8	9
0	—	0,1	0,3	0,4	0,5	0,6	0,8	0,9	1,0	1,2
1	1,29	1,42	1,54	1,67	1,80	1,93	2,06	2,19	2,32	2,45
2	2,58	2,70	2,83	2,96	3,09	3,22	3,35	3,48	3,61	3,73
3	3,86	3,99	4,12	4,25	4,38	4,51	4,64	4,76	4,89	5,02
4	5,15	5,28	5,41	5,54	5,67	5,79	5,92	6,05	6,18	6,31
5	6,44	6,57	6,70	6,82	6,95	7,08	7,21	7,34	7,47	7,60
6	7,73	7,86	7,98	8,11	8,24	8,37	8,50	8,63	8,76	8,89
7	9,01	9,14	9,27	9,40	9,53	9,66	9,79	9,92	10,04	10,17
8	10,30	10,43	10,56	10,69	10,82	10,95	11,07	11,20	11,33	11,46
9	11,59	11,72	11,85	11,98	12,10	12,23	12,36	12,49	12,62	12,75
10	12,88	13,01	13,14	13,26	13,39	13,52	13,65	13,78	13,91	14,04

Table 54

Conversion of Milligrams of N into Milligrams of NO_2^-

Whole mg of N	Tenths of a mg of N									
	0	1	2	3	4	5	6	7	8	9
0	—	0,3	0,7	1,0	1,3	1,6	2,0	2,3	2,6	3,0
1	3,28	3,61	3,94	4,27	4,60	4,93	5,25	5,58	5,91	6,24
2	6,57	6,90	7,23	7,55	7,88	8,21	8,54	8,87	9,20	9,52
3	9,85	10,18	10,51	10,84	11,17	11,49	11,82	12,15	12,48	12,81
4	13,14	13,47	13,79	14,12	14,45	14,78	15,11	15,44	15,78	16,09
5	16,42	16,75	17,08	17,41	17,74	18,06	18,33	18,72	19,05	19,38
6	19,71	20,03	20,36	20,69	21,02	21,35	21,68	22,00	22,33	22,66
7	22,99	23,32	23,65	23,98	24,30	24,63	24,96	25,29	25,62	25,95
8	26,28	26,60	26,93	27,26	27,59	27,92	28,25	28,57	28,90	29,23
9	29,56	29,89	30,22	30,54	30,87	31,20	31,53	31,86	32,19	32,52
10	32,84	33,16	33,49	33,82	34,15	34,48	34,80	35,13	35,46	35,79

Table 55

Conversion of Milligrams of N into Milligrams of NO_3^-

Whole mg of N	Tenths of a mg of N									
	0	1	2	3	4	5	6	7	8	9
0	—	0,4	0,9	1,3	1,8	2,2	2,7	3,1	3,5	4,0
1	4,43	4,87	5,31	5,75	6,20	6,64	7,08	7,52	7,97	8,41
2	8,85	9,30	9,74	10,18	10,62	11,07	11,51	11,95	12,39	12,84
3	13,28	13,72	14,16	14,61	15,05	15,49	15,94	16,38	16,82	17,26
4	17,71	18,15	18,59	19,03	19,48	19,92	20,36	20,80	21,25	21,69
5	22,13	22,58	23,02	23,46	23,90	24,35	24,79	25,23	25,67	26,12
6	26,56	27,00	27,44	27,89	28,33	28,77	29,22	29,66	30,10	30,54
7	30,99	31,43	31,87	32,31	32,76	33,20	33,64	34,08	34,53	34,97
8	35,41	35,85	36,30	36,74	37,18	37,63	38,07	38,51	38,95	39,40
9	39,84	40,28	40,72	41,17	41,61	42,05	42,49	42,94	43,38	43,82
10	44,27	44,71	45,15	45,59	46,04	46,48	46,92	47,36	47,81	48,25

Table 56

VII. Table for Converting Milligrams of NH_3 into Milligrams of NH_4^+

Whole mg of NH_3	Tenths of a mg of NH_3									
	0	1	2	3	4	5	6	7	8	9
0	—	0,1	0,2	0,3	0,4	0,5	0,6	0,7	0,8	1,0
1	1,06	1,17	1,27	1,38	1,48	1,59	1,69	1,80	1,91	2,01
2	2,12	2,22	2,33	2,44	2,54	2,65	2,75	2,86	2,97	3,07
3	3,18	3,28	3,39	3,50	3,60	3,71	3,81	3,92	4,02	4,13
4	4,24	4,34	4,45	4,55	4,66	4,77	4,87	4,98	5,08	5,19
5	5,30	5,40	5,51	5,61	5,72	5,83	5,93	6,04	6,14	6,25
6	6,36	6,46	6,57	6,67	6,78	6,88	6,99	7,10	7,20	7,31
7	7,41	7,52	7,63	7,73	7,84	7,94	8,05	8,16	8,26	8,37
8	8,47	8,58	8,69	8,79	8,90	9,00	9,11	9,22	9,32	9,43
9	9,53	9,64	9,74	9,85	9,96	10,06	10,17	10,27	10,38	10,49
10	10,59	10,70	10,80	10,91	11,02	11,12	11,23	11,33	11,44	11,55

Table 57

VIII . Table for Converting Oxidizability with Milligrams of $KMnO_4$

into Milligrams of O

Whole mg of $KMnO_4$	Tenths of a mg of $KMnO_4$									
	0	1	2	3	4	5	6	7	8	9
0	–	0,0	0,1	0,1	0,1	0,1	0,2	0,2	0,2	0,2
1	0,25	0,28	0,30	0,33	0,35	0,38	0,40	0,43	0,46	0,48
2	0,51	0,53	0,56	0,58	0,61	0,63	0,66	0,68	0,71	0,73
3	0,76	0,78	0,81	0,84	0,86	0,89	0,91	0,94	0,96	0,99
4	1,01	1,04	1,06	1,09	1,11	1,14	1,16	1,19	1,21	1,24
5	1,27	1,29	1,32	1,34	1,37	1,39	1,42	1,44	1,47	1,49
6	1,52	1,54	1,57	1,59	1,62	1,65	1,67	1,70	1,72	1,75
7	1,77	1,80	1,82	1,85	1,87	1,90	1,92	1,95	1,97	2,00
8	2,02	2,05	2,08	2,10	2,13	2,15	2,18	2,20	2,23	2,25
9	2,28	2,30	2,33	2,35	2,38	2,40	2,43	2,46	2,48	2,51
10	2,53	2,56	2,58	2,61	2,63	2,66	2,68	2,71	2,73	2,76

FACTORS FOR CONVERTING THE RESULTS OF WATER ANALYSIS FROM ONE FORM INTO ANOTHER

Table 58

Factors for Converting Milligrams of Certain Ions into Milligram—Equivalents

Ion	Equivalent weight of ion	Conversion factor	Ion	Equivalent weight of ion	Conversion factor
Li^+	6,940	0,144	S^{2-}	16,033	0,0624
Ba^{2+}	68,68	0,0146	HS^-	33,074	0,0302
Sr^{2+}	43,815	0,0228	$H_2PO_4^-$	96,991	0,0103
Cu^{2+}	31,77	0,0315	HPO_4^{2-}	47,991	0,0208
Pb^{2+}	103,605	0,00965	$HSiO_3$	77,098	0,0130
Zn^{2+}	32,69	0,0306	BO_2^-	42,82	0,0234

Table 59

Factor for Converting Some Oxides into Ions

Oxide	Ion	Conversion factor	Oxide	Ion	Conversion factor
Li_2O	Li^+	0,465	ZnO	Zn^{2+}	0,803
BaO	Ba^{2+}	0,896	As_2O_3	As	0,757
SrO	Sr^{2+}	0,846	P_2O_5	$H_2PO_4^-$	1,367
CuO	Cu^{2+}	0,799	P_2O_5	HPO_4^{2-}	1,352
PbO	Pb^{2+}	0,928			

Table 60

Factor for Converting Salts into Ions

Formula of salt	Molecular weight of salt	Conversion of factor	
		into cation	into anion
NaCl	58,448	0,3934	0,6066
Na$_2$SO$_4$	142,048	0,3237	0,6763
NaNO$_3$	84,999	0,2705	0,7295
NaHCO$_3$	84,010	0,2737	0,7263
Na$_2$CO$_3$	105,993	0,4338	0,5662
KCl	74,557	0,5244	0,4756
K$_2$SO$_4$	174,266	0,4487	0,5513
KNO$_3$	101,108	0,3867	0,6133
KHCO$_3$	100,119	0,3905	0,6095
K$_2$CO$_3$	138,211	0,5658	0,4342
CaCl$_2$	110,994	0,3611	0,6389
CaSO$_4$	136,146	0,2944	0,7056
Ca(NO$_3$)$_2$	164,096	0,2442	0,7558
Ca(HCO$_3$)$_2$	162,118	0,2472	0,7528
CaCO$_3$	100,091	0,4004	0,5996
MgCl$_2$	95,234	0,2554	0,7446
MgSO$_4$	120,386	0,2020	0,7980
Mg(NO$_3$)$_2$	148,336	0,1640	0,8360
Mg(HCO$_3$)$_2$	146,358	0,1662	0,8338
MgCO$_3$	84,331	0,2884	0,7116
Fe (HCO$_3$)$_2$	177,888	0,3140	0,6860

Conversion examples:
1. Given: 650.6 mg/liter of Na$_2$SO$_4$.
Na$^+$ = 0.3237 · 650.6 = 210.6 mg/liter; SO$_4^{2-}$ = 0.6763 · 650.6 = 440.0 mg/liter.
2. Given: 8.06 g/kg of MgCl$_2$.
Mg^{2+} = 0.2554 · 8.06 = 2.06 g/kg; Cl$^-$ = 0.7446 · 8.06 = 6.00 g/kg.

Table 61

Average Value of Activity Coefficient

Valance of ions	Ionic strength of solution μ 0	0,001	0,005	0,01	0,05	0,1
Monovalent	1	0,96	0,92	0,89	0,81	0,78
Divalent	1	0,86	0,72	0,63	0,44	0,33
Trivalent	1	0,73	0,51	0,39	0,15	0,08

The ionic strength of a solution μ is calculated from the formula

$$\mu = \frac{C_1 Z^2_1 + C_2 Z^2_2 + \ldots + C_n Z^2_n}{2},$$

where C_1, C_2, ..., C_n are the molar ion concentrations and Z_1, Z_2, ..., Z_n are the corresponding valences of these ions.

It is more convenient to calculate the ionic strength of natural water by the following formula:

$$\mu = \frac{\sum_r u_1 + 2\sum_r u_2 + 3\sum_r u_3}{2 \cdot 1000},$$

where $\sum_r u_1$, $\sum_r u_2$, and $\sum_r u_3$ are the sums of the milligram-equivalents of the mono-, di-, and trivalent ions, respectively, contained in the water examined.

Calculation Example. Found in the water (in milligram-equivalents per liter):

$$Ca^{2+} - 2.0; \quad Mg^{2+} - 1.5; \quad Na^+ - 8.0;$$

$$Cl^- - 3.0; \quad SO_4^{2-} - 4.5; \quad \text{and} \quad HCO_3^- - 4.0$$

$$\mu = \frac{(_r Na^+ + _r Cl^- + _r HCO_3^-) + 2(_r Ca^{2+} + _r Mg^{2+} + _r SO_4^{2-})}{2 \cdot 1000} =$$

$$= \frac{(8.0 + 3.0 + 4.0) + 2(2.0 + 1.5 + 4.5)}{2 \cdot 1000} = 0.0155.$$

The average value of the activity coefficient f_{av} is found in Table 61 from the calculated ionic strength μ. If the calculated value of μ does not correspond to the figures given in the table, the average activity coefficient is found by interpolation. In the example given above ($\mu = 0.0155$) the average activity coefficient f_{av} will equal 0.88 for monovalent ions, 0.60 for divalent ions, and 0.36 for trivalent ions.

NOMOGRAM FOR CALCULATING PERCENT-EQUIVALENTS

The nomogram (Fig. 1) makes it possible to calculate the percent-equivalents for any mineral content, accurate to whole numbers.

An example of a calculation, the key to the nomogram, is also given in Fig. 1, and as it is simple, it requires no explanation.

A more complicated example of a calculation is given below.

Anions found in water: $HCO_3^- - 8.32$ meq/liter, $SO_4^{2-} - 2.62$ meq/liter, $Cl^- - 17.43$ meq/liter, $NO_3^- - 0.40$ meq/liter. The sum of the milligram-equivalents of the anions is 28.77. This value does not appear on line I (the sum of the milligram-equivalents); therefore we find on this line the point corresponding to a tenth of this value, 2.88 meq, connect it to points on line II (milligram-equivalents) corresponding to milligram-equivalents found by analysis, and read off the result on line III (percent-equivalent). Since the sum of the milligram-equivalents was divided by 10, the values found must be divided by the same factor. Depending on the limits of the nomogram, this reduction may be made either on line I or on line II. Thus, in the given example, for HCO_3^- and Cl^- (8.32 and 17.43 meq) the reduction is made on line II (the points 0.83 and 1.74 meq) and by joining the point on line I (2.88 meq) to the points on line II (0.83 and 1.74 meq) with straight lines and extending them to intersect line III, we can read off the results: 29 and 61%-equiv. For SO_4^{2-} and NO_3^- the reduction is made on line III (the results obtained, 93 and 14%-equiv, are divided by 10 and rounded off to whole numbers; the final results are 9 and 1%-equiv, respectively).

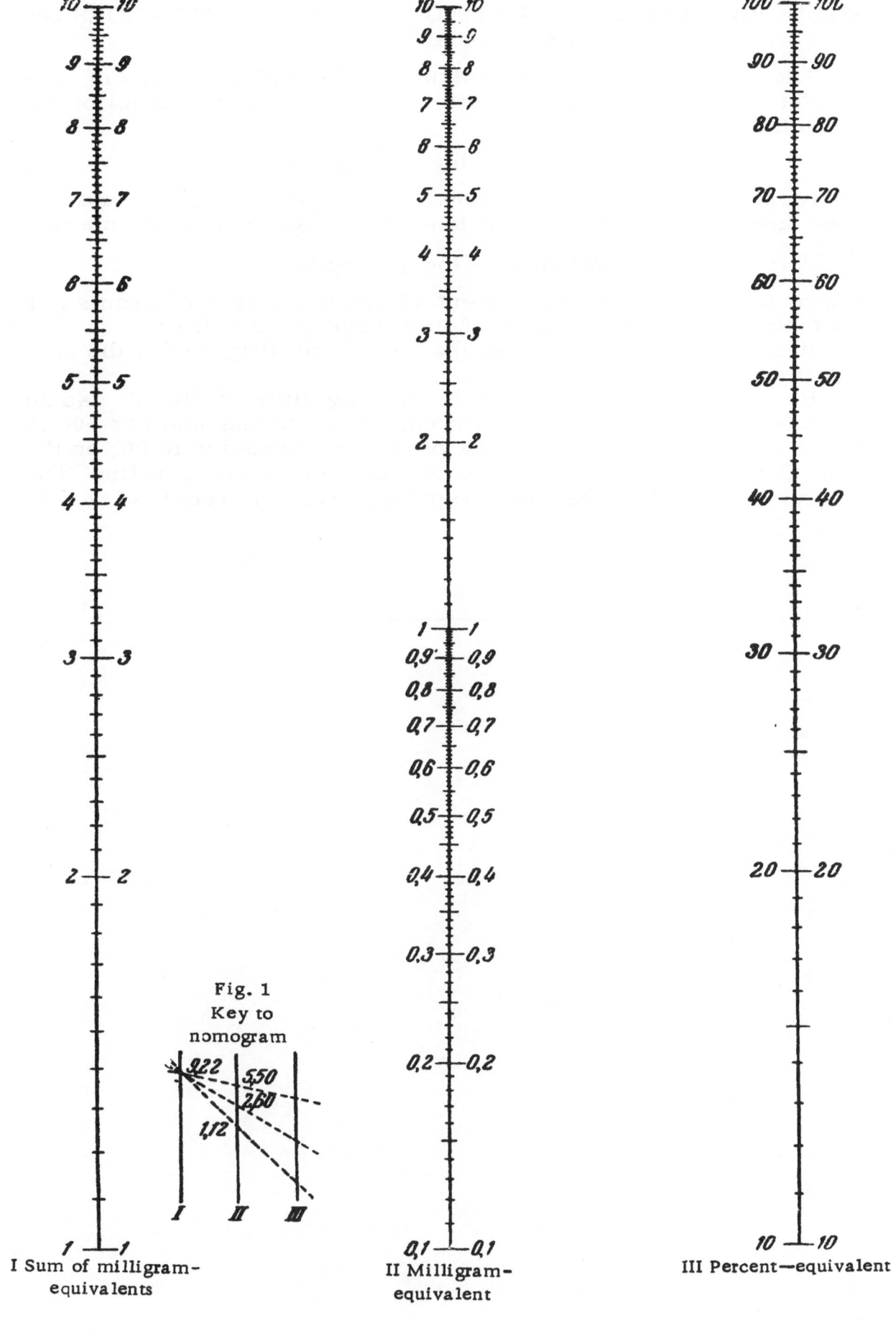

Fig. 1
Key to
nomogram

I Sum of milligram-
equivalents

II Milligram-
equivalent

III Percent—equivalent

NOMOGRAM FOR CALCULATING pH FROM GIVEN VALUES OF FREE CO_2 AND HCO_3^-

The pH of water may be calculated from given values of free carbon dioxide and bicarbonate ion from one of the following formulas:

$$pH = 8.16 - \log a + \log n, \tag{1}$$

where a is the free CO_2 content of the water in milligrams per liter and n is the HCO_3^- content in milligram-equivalents per liter; or

$$pH = 6.38 - \log a + \log b, \tag{2}$$

where a is the free CO_2 content of the water in milligrams per liter and b is the HCO_3^- content in milligrams per liter.

It is much simpler to use the nomogram (Fig. 2) for the calculation.

Example. Found in water: 400 mg/liter of HCO_3^- and 30 mg/liter of CO_2. The point corresponding to the number 400 is found on the HCO_3^- line and the point corresponding to 30, on the free CO_2 line, and these points are joined by a straight line. The intersection of this line with the pH line gives the required result, 7.5.

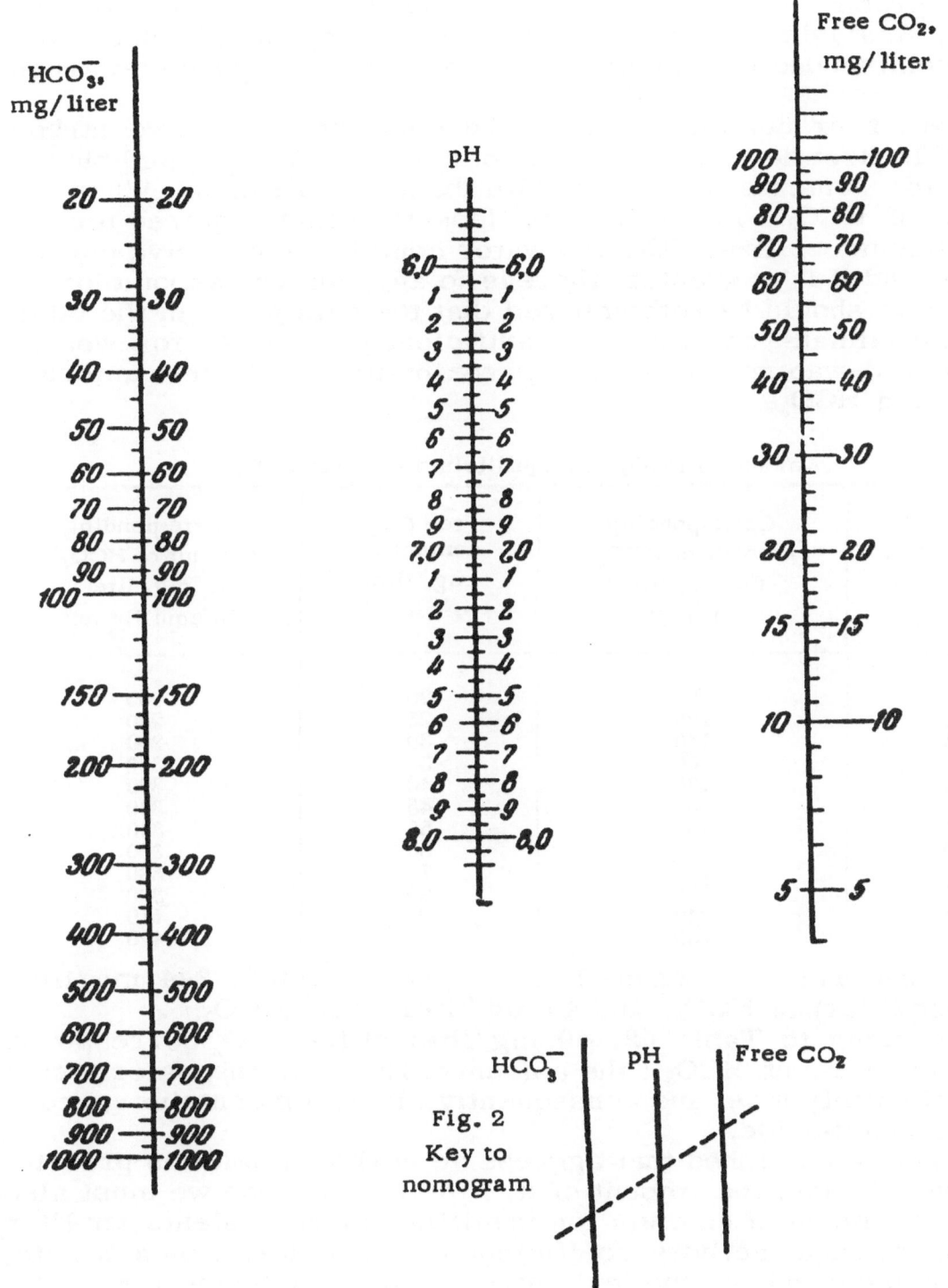

HCO₃⁻, mg/liter

HCO_3^-, mg/liter

20 — 20
30 — 30
40 — 40
50 — 50
60 — 60
70 — 70
80 — 80
90 — 90
100 — 100
150 — 150
200 — 200
300 — 300
400 — 400
500 — 500
600 — 600
700 — 700
800 — 800
900 — 900
1000 — 1000

pH

6,0 — 6,0
1 — 1
2 — 2
3 — 3
4 — 4
5 — 5
6 — 6
7 — 7
8 — 8
9 — 9
7,0 — 7,0
1 — 1
2 — 2
3 — 3
4 — 4
5 — 5
6 — 6
7 — 7
8 — 8
9 — 9
8,0 — 8,0

Free CO₂, mg/liter

CO_2, mg/liter

100 — 100
90 — 90
80 — 80
70 — 70
60 — 60
50 — 50
40 — 40
30 — 30
20 — 20
15 — 15
10 — 10
5 — 5

HCO_3^- pH   Free CO_2

Fig. 2
Key to
nomogram

CALCULATION OF AGGRESSIVE CARBON DIOXIDE

The graphs (Figs. 3 and 4) developed by F. F. Laptev (F. F. Laptev, Analysis of Water [in Russian], Gosgeoltekhizdat, 1955) are recommended for calculating the amount of aggressive carbon dioxide.

Whether or not the water tested contains aggressive carbon dioxide is first determined by comparing the free CO_2 and bicarbonate ion contents found in it with the aid of Table 62. When the amount of HCO_3^- is found to be less than that required for the given amount of free CO_2, the water contains aggressive carbon dioxide, and if it is greater, there is no aggressive carbon dioxide present. It should be remembered that the data given in the table are approximate and are only sufficiently accurate for weakly mineralized waters containing approximately equivalent amounts of Ca^{2+} and HCO_3^-.

Table 62

Amount of HCO_3^- in Equilibrium with Free CO_2

Free CO_2 content in mg/liter	Corresponding amount of HCO_3^- (in mg/liter) in equilibrium	Free CO_2 content in mg/liter	Corresponding amount of HCO_3^- (in mg/liter) in equilibrium
1	80	20	260
2	120	25	280
3	140	30	290
4	160	35	300
5	170	40	320
6	180	45	330
8	200	50	340
10	210	60	360
12	220	70	380
14	230	80	390
16	240	90	410
18	250	100	420

Calculation Example. Found in water: 244 mg/liter (4.0 meq/liter) of HCO_3^- and 40 mg/liter of free CO_2.

According to Table 62, 40 mg/liter of free CO_2 corresponds to 320 mg/liter of HCO_3^-; the amount of HCO_3^- in the water tested is considerably less, and consequently the water contains aggressive carbon dioxide.

Having established that aggressive carbon dioxide is present, we now calculate the amount of it. For this purpose we must also know the amount of calcium ion in milligram-equivalents per liter and the average activity coefficient of the calcium bicarbonate, f_{av}, which depends on the ionic strength of the given water.

The relation between the average activity coefficient of calcium bicarbonate, f_{av}, and the ionic strength μ is given in Table 63 (values of the activity coefficients of mono- and divalent ions from Table 61 were used to draw up this table).

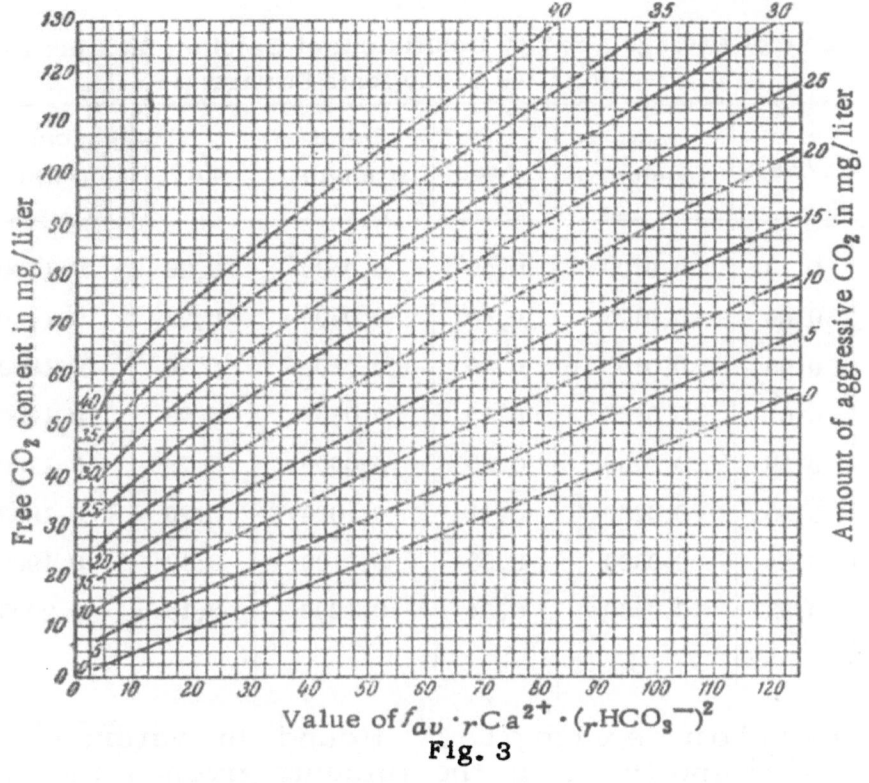

Fig. 3

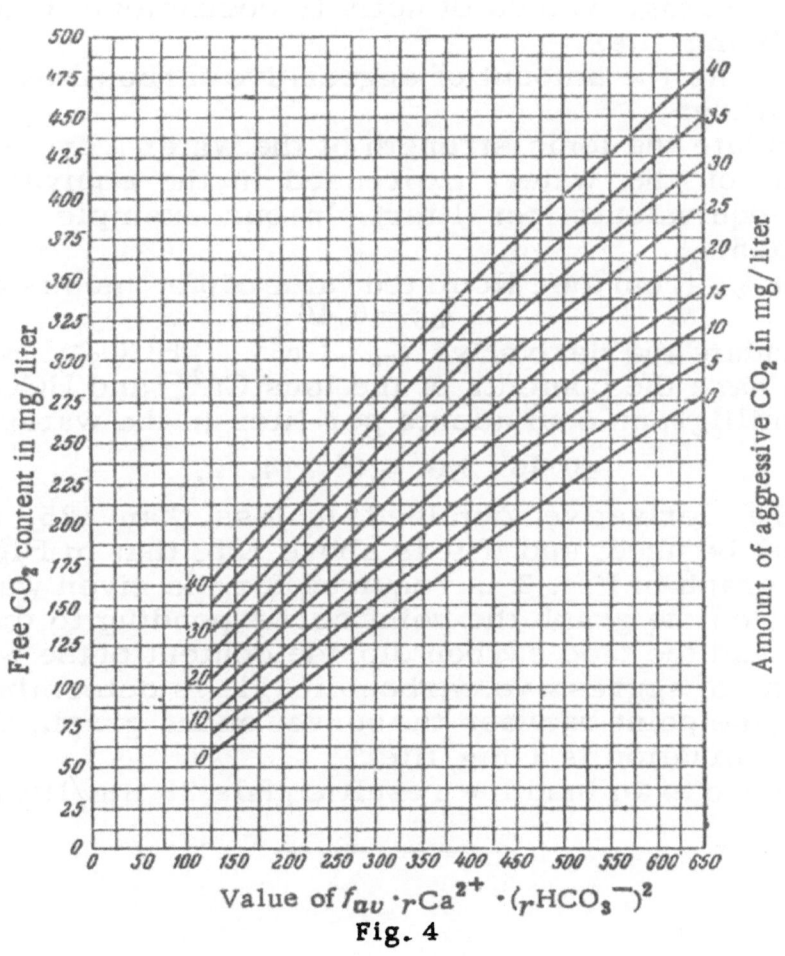

Fig. 4

81

Table 63

Value of Average Activity Coefficient of Calcium Bicarbonate f_{av}
Relative to the Ionic Strength μ

Ionic strength μ	f_{av}	Ionic strength μ	f_{av}	Ionic strength	f_{av}	Ionic strength μ	f_{av}
0,0025	0,73	0,0100	0,50	0,0500	0,29	0,1300	0,17
0,0030	0,69	0,0125	0,48	0,0600	0,27	0,1400	0,16
0,0040	0,65	0,0150	0,46	0,0700	0,26	0,1500	0,15
0,0050	0,61	0,0175	0,45	0,0800	0,23	0,1600	0,15
0,0060	0,59	0,0200	0,43	0,0900	0,22	0,1700	0,14
0,0070	0,56	0.0250	0,41	0,1000	0,20	0,1800	0,13
0,0080	0,54	0,0300	0,38	0,1100	0,19	0,1900	0,13
0,0090	0,53	0,0400	0,34	0,1200	0,18	0,2000	0,12

Calculation example. Found in water: Ca^{2+}, Mg^{2+}, Na^+, Cl^-, SO_4^{2+}, and HCO_3^- in the amounts given in the example of finding the average values of activity coefficients (see p. 75) and free CO_2, 40 mg/liter.

To calculate the amount of aggressive carbon dioxide, we must do the following:

1. Calculate the ionic strength of the water μ from the results of analysis of the water, expressed in the equivalent form (in milligram-equivalents per liter). In our example (see p. 75) it equals 0.0155.

2. Find f_{av} from the calculated value of μ by means of Table 63:
$$f_{av} = 0.46.$$

3. Calculate the derivative $f_{av} \cdot {}_r Ca^{2+} \cdot ({}_r HCO_3^-)^2$, where ${}_r Ca^{2+}$ and ${}_r HCO_3^-$ are the contents of the ions Ca^{2+} and HCO_3^-, respectively, in milligram-equivalents per liter in the water:

$$0.46 \cdot 2.0 \cdot 4.0^2 = 14.72.$$

4. If the derivative obtained is less than 125, the graph in Fig. 3 must be used, and if it is above 125, that in Fig. 4 must be used. The graph in Fig. 3 is required for the given water.

5. Find on the graph the point corresponding to the calculated derivative and the free carbon dioxide content of the water tested. The amount of aggressive carbon dioxide is determined from the position of the point between the curves on the graph; the accuracy of the determination is 1 mg/liter.

In the case examined the result equals 23 mg/liter of $CO_{2\,agg}$.

Table 64

Relation between Different Forms of Weak Acids in Natural Waters at Different pH Values and Ionic Strengths (in molar percent)

The numerator gives the percent content at $\mu = 0$, and the denominator that at $\mu = 0.1$

Components	pH value					
	5	6	7	8	9	10
Carbon dioxide						
$[CO_2]$	97/96	77/72	25/20	3/3	0/0	0/0
$[HCO_3-]$	3/4	23/28	75/80	96/96	95/89	67/46
$[CO_3^{2-}]$	0/0	0/0	0/0	1/1	5/11	33/54
Hydrogen sulfide						
$[H_2S]$	99/99	95/93	64/58	15/12	2/1	0/0
$[HS-]$	1/1	5/7	36/42	85/88	98/99	100/100
$[S^{2-}]$	0/0	0/0	0/0	0/0	0/0	0/0
Phosphoric acid						
$[H_3PO_4]$	0/0	0/0	0/0	0/0	0/0	0/0
$[H_2PO_4^-]$	99/99	94/87	62/41	14/6	2/1	0/0
$[HPO_4^{2-}]$	1/1	6/13	38/59	86/94	98/99	100/99
$[PO_4^{3-}]$	0/0	0/0	0/0	0/0	0/0	0/1
Silicic acid						
$[H_2SiO_3]$	100/100	100/100	99/99	91/89	50/44	9/7
$[HSiO_3-]$	0/0	0/0	1/1	9/11	50/56	91/93
$[SiO_3^{2-}]$	0/0	0/0	0/0	0/0	0/0	0/0

N o t e.   The following ionization constants were used for the calculations:

for carbon dioxide $K_1 = 3 \cdot 10^{-7}$; $K_2 = 5 \cdot 10^{-11}$

for hydrogen sulfide $K_1 = 5.7 \cdot 10^{-8}$; $K_2 = 1.2 \cdot 10^{-15}$

for phosphoric acid $K_1 = 7.5 \cdot 10^{-3}$; $K_2 = 6.2 \cdot 10^{-8}$; $K_3 = 2.2 \cdot 10^{-13}$

for silicic acid $K_1 = 10^{-9}$; $K_2 = 10^{-13}$

The ionization constants of silicic acid are approximate and consequently the relations between the different forms of silicic acid given are approximate

Table 65

Factors for Converting Different Forms of Expressing Aqueous Solution Concentrations into Milligrams and Milligram-Equivalents

Form of expressing aqueous solution concentration	Conversion factor	
	into mg/liter	into meq/liter
Weight percent (%).	$d \cdot 10\,000$	
Parts per thousand by weight ($^0/_{00}$)	$d \cdot 1000$	
Parts per million by weight (ppm)	d	
Grams per liter (g/liter)	1000	
Gamma or micrograms per liter (γ/liter, μg/liter) or milligrams per cubic meter (mg/m³).	0.001	
Molar solutions (M)		$A \cdot 1000$
Normal solutions (N).		1000
Microgram-equivalents per liter (μg-equiv/liter)		0.001
Milligram-equivalents per 100 g of solution. . . .		$d \cdot 10$

N o t e. d is the specific gravity of the solution examined and A is the basicity of the substance (or for an ion, its valence).

The tables given in the present work may be used for interconversion of milligram-equivalents per liter.

C a l c u l a t i o n   E x a m p l e. A brine has a salt concentration of 8.50% and a specific gravity of 1.060.

The salt content in milligrams per liter equals $8.50 \cdot 1.060 \cdot 10,000 = 90,100$.

International Atomic Weights

Name of element	Atomic number	Symbol	Atomic weight	Name of element	Atomic number	Symbol	Atomic weight
Actinium	89	Ac	227	Mercury	80	Hg	200.61
Aluminum	13	Al	26.98*	Molybdenum	42	Mo	95.95
Americium	95	Am	[243]*	Neodymium	60	Nd	144.27
Antimony	51	Sb	121.76	Neon	10	Ne	20.183
Argon	18	A	39.944	Neptunium	93	Np	[237]
Arsenic	33	As	74.91	Nickel	28	Ni	58.69
Astatine	85	At	[210]	Niobium	41	Nb	92.91
Barium	56	Ba	137.36	Nitrogen	7	N	14.008
Berkelium	97	Bk	[245]	Osmium	76	Os	190.2
Beryllium	4	Be	9.013	Oxygen	8	O	16
Bismuth	83	Bi	209.00	Palladium	46	Pd	106.7
Boron	5	B	10.82	Phosphorus	15	P	30.975
Bromine	35	Br	79.916	Platinum	78	Pt	195.23
Cadmium	48	Cd	112.41	Plutonium	94	Pu	[242]
Calcium	20	Ca	40.08	Polonium	84	Po	210
Californium	98	Cf	[248]	Potassium	19	K	39.100
Carbon	6	C	12.011	Praseodymium	59	Pr	140.92
Cerium	58	Ce	140.13	Promethium	61	Pm	[145]
Cesium	55	Cs	132.91	Protactinium	91	Pa	231
Chlorine	17	Cl	35.457	Radium	88	Ra	226.05
Chromium	24	Cr	52.01	Radon	86	Rn	222
Cobalt	27	Co	58.94	Rhenium	73	Re	186.31
Copper	29	Cu	63.54	Rhodium	45	Rh	102.91

Element	Number	Symbol	Atomic Weight	Element	Number	Symbol	Atomic Weight
Curium	96	Cm	[245]	Rubidium	37	Rb	85.48
Dysprosium	66	Dy	162.46	Ruthenium	44	Ru	101.1
Einsteinium	99	En	[253]	Samarium	62	Sm	150.43
Erbium	68	Er	167.2	Scandium	21	Sc	44.96
Europium	63	Eu	152.0	Selenium	34	Se	78.96
Fermium	100	Fm	[255]	Silicon	14	Si	28.09
Fluorine	9	F	19.00	Silver	47	Ag	107.880
Francium	87	Fr	[223]	Sodium	11	Na	22.991
Gadolinium	64	Gd	156.9	Strontium	38	Sr	87.63
Gallium	31	Ga	69.72	Sulfur	16	S	32.066
Germanium	32	Ge	72.60	Tantalum	73	Ta	180.95
Gold	79	Au	197.0	Technetium	43	Tc	[99]
Hafnium	72	Hf	178.6	Tellurium	52	Te	127.61
Helium	2	He	4.003	Terbium	65	Tb	158.93
Holmium	67	Ho	164.94	Thallium	81	Tl	204.39
Hydrogen	1	H	1.0080	Thorium	90	Th	232.05
Indium	49	In	114.76	Thulium	69	Tm	168.94
Iodine	53	I	126.91	Tin	50	Sn	118.70
Iridium	77	Ir	192.2	Titanium	22	Ti	47.90
Iron	26	Fe	55.85	Tungsten	74	W	183.92
Krypton	36	Kr	83.80	Uranium	92	U	238.07
Lanthanum	57	La	138.92	Vanadium	23	V	50.95
Lead	82	Pb	207.21	Xenon	54	Xe	131.3
Lithium	3	Li	6.940	Ytterbium	70	Yb	173.04
Lutecium	71	Lu	174.99	Yttrium	39	Y	88.92
Magnesium	12	Mg	24.32	Zinc	30	Zn	65.38
Manganese	25	Mn	54.94	Zirconium	40	Zr	91.22
Mendelevium	101	Mv	[256]				

* The numbers in brackets indicate the mass number of the isotope with the longest half-life.